MW00510953

For more information:
quietriverpress.com

Practical Home Theater

A Guide to Video and Audio Systems

2011 EDITION

by Mark Fleischmann

QUIET RIVER PRESS

NEW YORK

© Copyright 2010 Mark Fleischmann. All rights reserved.

No part of this book may be reproduced, stored in a retrieval system, or transmitted by any means, electronic, mechanical, photocopying, recording, or otherwise, without written permission from the author.

ISBN 13: 978-1-932732-12-2
ISBN 10: 1-932732-12-8

This book is printed on acid-free paper.

quietriverpress.com

Dedicated to my mother, Lucille Fleischmann, who packed me into her pink-finned Plymouth and conveyed me to the Middlesex and Bound Brook, New Jersey public libraries when I was small and impressionable. This led to a fascination with books and a career pushing words in newspaper, magazine, online, and now book publishing. Thanks, Mom.

Table of Contents

Introduction

What is home theater?

I like home theater because it is lots of fun. Thank you, ladies and gentlemen, goodnight!

Welcome to home theater, square one. Though I fancy myself a home theater critic, credit for coining the term goes to Sam Runco, the projector king. I define home theater in three ever more specific ways: as a state of mind, as a room, and finally as a system playing in a room to a small, thrilled audience.

What is the home theater state of mind? It's the rush of emotional involvement that comes with a close-up shot on a big screen. It's the aesthetic thrill of hearing a rich orchestral score through an enveloping array of speakers. It's the quickening of the pulse that follows a gut-thumping subwoofer-borne explosion. It's about having more fun at home than you ever did at the movies. The movies have come home.

Where in the home does a home theater go? Wherever it can. Those with a spare bedroom or basement to devote to the hobby can end up with a luxurious screening room rivaling that of the early 20th century movie moguls. Then again, a home theater system may be a nonintrusive part of a multi-purpose room—until the lights go out. Then it transforms your living space, as a beam of brightness hits the wall and sound fills the room.

What are the elements of a *practical* home theater system? I'm talking about a movie *and* music system that takes best advantage of current technology but remains within a realistic budget. It starts with a big-screen display, coupled with a surround sound system. Movies, music, and other programming arrive via disc, tape, download, streaming, server, satellite, cable, or over the air—and fill in the past or future format of your choice. The right cables, remote, and other accessories tie the system together.

According to the Consumer Electronics Association, a trade group representing manufacturers, about 33 percent of American households had home theater systems as of July 2009, as compared with 6 percent in 1998, and there was no further change in 2010. Given the universal love of movies and music, that leaves a lot of room for growth. Clearly many of us find the technology intimidating. The purpose of this book is to enable the reader to become familiar and comfortable with the technology behind home theater systems.

The book aims to be comprehensive. As a result, reading it from start to finish may be rough going (as my psychiatrist has noted rather pointedly). It's not designed to be something you'd take to the beach. However, it does work well as an answer book. Read the relevant chapter when you're about to buy something—the table of contents will send you to the right place. Or dip into the book when you have a question, using the index as your guide.

What is big-screen television?

A big picture is one of two key elements in a home theater system, in addition to surround sound.

Any of several kinds of big-screen television may form the basis of a home theater system. Which one is right for you is a question of room size as well as videophile taste.

For bedroom systems and small rooms in general, the tube TV—or **direct-view TV,** to use industry parlance—still survives in many homes, though I don't recommend buying a new one. Tube TVs can be analog or digital; only the latter are now sold. Analog tube TVs fed with over-the-air signals need a set-top convertor now that the transition from analog to digital broadcasting is complete. Tube TVs have been phased out as hipper consumers increasingly opt for **flat panel** displays using **LCD** technology. A tube TV, even a digital one, is no longer a good investment. Flat panels are more high-def-worthy and far more elegant.

For medium-sized rooms, larger LCDs are available, and **plasma** display panels become a viable alternative. LCDs and plasmas are available in different ranges of size, with plasmas tending to be larger, but there is some overlap. They work in dramatically different ways. LCDs create illuminated dots onscreen by combining a backlight with liquid crystals that open and close like light valves, whereas plasmas are big glass sandwiches dotted with gas-filled neon tubes that light up when stimulated by current.

Another option for a medium-sized room is a **rear-projection TV**. Though all but a few manufacturers have exited the category, it still offers good value—that is, the most inches of picture per dollar. Rear-projectors with **CRTs** (**cathode ray tubes**) offer filmlike picture quality in (unfortunately) the biggest chassis. That is why they have all but disappeared, supplanted by less bulky **microdisplay** rear-projectors that replace the bulky tubes with imaging chips. These slimmer projectors use either **DLP** or **LCD** technology.

For the biggest rooms, the biggest pictures come with two-piece **front-projection TVs** that splash the picture on a screen or even a blank white wall. Bulky projectors that use big CRTs provide the best black level, and for that reason are still used by a few cutting-edge videophiles. But most people now prefer projectors using solid-state DLP or LCD technology, allowing a smaller projector to produce a bright, colorful picture.

Any of these video technologies has the potential to display **digital television** (**DTV**), known in its best form as **high-definition television** (**HDTV**). However, the benefits of HDTV are best enjoyed on a large screen—the larger, the better. Big screens show off the detail and sharpness of HDTV *and* deliver greater emotional impact in general. While your eyes feast, your emotions get pulled into the story.

As mentioned above, 2009 marked a milestone in the history of television, with analog broadcasting replaced by digital broadcasting on June 12 of that year. You needn't worry about this if you own a DTV with digital tuner, or get video delivered by cable, satellite, or telco. That would include most readers of this book. But if you—or someone you know, possibly someone elderly or poor—depends on an analog TV with over-the-air signals, a set-top box will be necessary to convert the new digital broadcast signals to an analog format for the old set. See dtvtransition.org for details.

The "Television" chapter will discuss big-screen display technologies, and the basics of digital television, including resolution, screen shape, 3D, tuners, and connections—finishing with advice on how to shop for a DTV.

What is surround sound?

Surround sound is the second key element of a home-theater system, after big-screen television.

A basic home surround system feeds a **5.1-channel** array of three front left/center/right speakers, two surround speakers on the side walls, and a subwoofer (the ".1"). In larger rooms, a **6.1-** or **7.1-channel** array adds one or two extra speakers against the back wall. Alternatively, the sixth and seventh channels can be used to drive front-height or front-width channels. Each of the speakers in a 5.1 array specializes in a different part of the movie experience. The front-center channel handles dialogue, while the surrounds provide ambience and effects, and the sub handles bass. To feed these speakers you would use either a one-box **surround receiver** or a two-box combination of **surround preamp-processor** and **multichannel power amp**.

Movies produced over the past 30 years or so have surround sound buried in their soundtracks. The surround formats that bring it home are licensed by Dolby Laboratories and DTS. There are roughly three generations of them.

The original **Dolby Surround** is most likely to be found embedded into old videocassettes. It is decoded in home gear by **Dolby Pro-Logic**, **DPLII**, **DPLIIx**, or **DPLIIz**. The second generation, embedded into DVDs is **Dolby Digital 5.1**, whose main competitor is **DTS 5.1**. There are also 6.1- and 7.1-channel versions, **Dolby Digital EX** and **DTS-ES**. The third and latest generation of Dolby formats includes **Dolby Digital Plus** and lossless **Dolby TrueHD**, while the next generation of DTS formats includes **DTS-HD High Resolution Audio** and lossless **DTS-HD Master Audio**. The lossless formats are dominant on Blu-ray discs.

THX originally got started not as an alternative surround format but as a certification program that covers public movie theaters as well as every component in a home theater system including receivers, speakers, displays, cables, and software. Think of it as a high-end option that coexists with the Dolby and DTS surround standards. THX also offers a useful low-volume listening mode in Ultra2 Plus and Select2 Plus certified products.

Audyssey, the newest major player in surround, licenses a variety of technologies including auto setup, room correction, low-volume listening modes, and height/width listening modes.

The chapter on "Surround Sound" will tell everything you need to know about home theater speakers, receivers, preamp-processors, power amps, and the surround standards underlying them.

What's the cost?

Cost is an issue for readers of a book entitled *Practical Home Theater*. Regardless of what you buy, you may be about to lay out a lot of money for the pleasure of enjoying big-screen television and surround sound in your home. To make yourself an informed consumer, you will need to look at a broad selection of video and audio products—the best and brightest, as well as the cheapest and ugliest, and what's in between.

People are always asking me what they should buy. This makes me nervous and should make you even more so—it's always better to understand what you're buying than to depend on some self-appointed guru. However, to get you started, I've named names in pictures and captions strewn throughout the book. Don't take these suggestions too literally. The products featured here are not necessarily *things you should run out and buy*. Rather, they are *things you should look at* to become better informed. Are the prices prohibitive? In many cases, yes, of course, but it's still helpful to learn what top-drawer products and critic's picks offer as a reference for evaluating other products. Only then can you gauge how much performance, and how many features, you need in your own system.

In the final analysis, how much should you spend for a home theater system? Other critics would pull dollar figures out of the air or recommend that you spend a certain proportion on speakers. Sorry, I refuse to play this game. But here are some tentative answers:

Don't spend more than you can afford. Don't spend enough to prevent you from buying or renting movies and music to play on your system. Whatever you do spend should leave enough room for a good life, domestic harmony, and a nice bottle of wine with dinner. After dinner is when the fun begins.

Television

Welcome to the heart of home theater. Big-screen TV is half of the home theater equation. For some folks—not you blithe guilt-free consumers, but the remainder—buying a big-screen TV means breaking taboos. After all, isn't it the height of conspicuous consumption? Why buy a big expensive TV when stores are full of smaller and cheaper ones? And won't it take up an outrageous amount of room?

The answers: If you really want to enjoy movies at maximum emotional impact—and who doesn't?—a video display that serves as the centerpiece of a home theater system has to be large enough to dominate your senses. That's the only way to suspend disbelief and immerse yourself in the story. A big screen doesn't have to cost much, or take up much room. Flat panels and front-projection systems can produce a huge picture with literally no footprint.

To get the best value out of a big-screen purchase, a consumer must become well-informed. Besides variations in screen size, digital TVs also vary in the way they handle different video formats, screen shape, and their need for outboard equipment to receive broadcast, cable, or satellite TV. Subsequent chapters will cover all these issues. But first, let's talk about display technologies.

Big screens

The introduction discussed various kinds of television sets and video displays in terms of scale: for small rooms, LCD and direct-view TVs; for medium-sized rooms, bigger LCDs, plasmas, or rear-projection TVs (DLP, LCD, or CRT); and for the largest rooms, front-projection systems (using those same technologies). This chapter will change focus from scale to display technology, though moving in roughly the same direction—from small to large.

Digital vs. analog television

Digital television, or DTV, is different from analog television. The picture is cleaner and can be sharper—much sharper. Colors, especially bright red, look more natural because they're more accurate (or can be, anyway). Paradoxically, the best digital television sets are also noteworthy for the way they handle analog television, which still survives in analog cable and satellite channels and other legacy sources.

Even if you forget everything else you read in this book, remember this bit of advice: *Don't buy an analog TV* (even if you somehow find one for sale. And if you own one, consider replacing it. True, HDTV is an evolving technology—but it has evolved far enough to deserve your dollars. In terms of model-years, HDTV has been around for more than a decade. There's plenty of HD programming available via broadcast, cable, and satellite. The supply is being enlarged by the entry of the telephone companies into the video-delivery business, plus the high-definition Blu-ray disc format. High-def downloads and streaming are small but growing video program sources.

So what are your fellow consumers actually doing? Digital TV has taken over the market, having surpassed analog TV sales for the first time in 2005, according to the Consumer Electronics Association. By the end of 2009, according to CEA, 77 percent of U.S households had DTVs and 63 percent had HDTVs. The average LCD set sold for $560 in 2009 (projected to fall to $417 in 2010) and the average plasma for $955 in 2009 (up slightly, but projected to fall to $875 in 2010). No matter how little you have to spend, don't use an analog TV in a home theater system. The practical home theater buff is one who buys *good* digital at a price the household can afford.

HDTV's sharper picture can benefit you in several ways. Its greater resolution lets you sit closer to the screen, magnifying the emotional impact of a good movie. Or you could keep viewing distance the same but have a larger screen. If the room's arrangement is comfortably settled, keep screen size and viewing distance the same, while enjoying higher performance in a screen the same size as your old TV. Whether you sit closer, increase the screen size, or just enjoy higher performance, HDTV should make a difference in how well your home theater system delivers movies.

Direct-view tube TV

Most major TV brands have exited this category in the North American market, though tube TVs are still popular in some overseas markets, and you still might find a few models from small brands. The following information is therefore mainly historical.

The industry refers to TVs with a single large picture tube as **direct view** to distinguish them from projectors that also use tubes. Literally speaking, you directly view the front of the tube, which has electrons flung at its surface from the narrow **yoke** at the back of the tube. A *digital* tube TV musters high brightness and contrast and works well in multi-purpose rooms where it's not practical to keep the lights off. HD-capable versions have become increasingly affordable. There are also SDTVs (standard-definition) at even lower prices. Analog direct-view TVs are obsolete.

Basic tube TVs can be as small as 5 inches or as big as 40, measured diagonally. The minimum for an absorbing movie presentation should be 27 inches (nonwidescreen) or 34 inches (widescreen). Step up to higher sizes if you want more of a good thing—but for something really big, step up to flat panels and projectors.

Some picture enhancements owe more to marketing than technology. TV makers have stretched the laws of physics to see who can make the flattest **flat tube** before the sealed piece of glass caves in upon itself. A flatter tube will pick up fewer reflections from lamps, windows, and other bright objects in the room.

High contrast, or black tinted, faceplates and tubes do improve contrast. However, there's a tradeoff. They also lose a little light output and drive the tube harder. That means the tube won't last as long. You

3

can extend the life of any TV tube—and save yourself a bit of eye-strain—by lowering the contrast and brightness settings. Reducing ambient light will make the lowered settings seem more natural.

A tube constructed with an **Invar shadow mask**, a porous exotic-metal plate behind the picture tube, produces a finer-grained picture by more accurately guiding the movement of the electrons through its tiny holes to the red, green, and blue phosphors. Look for this feature on better tube displays.

The shadow mask is what limits the resolution of direct-view HDTVs. As the holes in the mask are made smaller and more numerous, it becomes more difficult for light to get through them. As a result, it's impossible for a direct-view HDTV to produce all the resolution of HDTV with all the light output consumers expect in a tube TV. No consumer-level direct-view TV can display all the scan lines in a 1080i HDTV test pattern.

Flat-panel display technologies

Among the most popular new displays are flat panel sets using either liquid crystal technology (for sets of any size) or gas plasma technology (for midsized to larger ones). LCD TVs in smaller sizes are now preferred to direct-view sets. Plasma and larger LCD panels compete with larger rear-projection sets.

Plasma, LCD, and DLP are referred to as **fixed pixel** displays because their grids of vertical and horizontal pixels are fixed. CRT-based displays, in contrast, do not have fixed pixel grids.

Flat-panel displays: LCD

LCD stands for **liquid crystal display**. Flat-panel sets using LCD technology range from 2.5 inches (in an iPod video player) to 108 inches (the current record holder is Sharp). The best values are in the medium-sized 32- to 42-inch range. When you get into the forties, plasma becomes a viable alternative. LCD TVs can be high-definition or standard-definition—make sure to buy the HD version except for casual viewing.

Most larger LCD sets are HDTV-worthy. They can sit on a pedestal or mount to the wall—in fact, at about half the weight of same-sized plasma displays, LCD flat panels are easier to mount. Note that not all LCD TVs are suitable for internet surfing. While your laptop uses an LCD, you can't assume that all LCDs are PC-compatible. LCDs designed for computer use may not have video inputs for television formats. However, mainstream TV manufacturers now include broadband connectivity which supports a variety of networked media features. Many new companies—some of which have never before sold television sets in the United States—have enter the LCD TV market in recent years. It is the fastest-growing category in television.

How do LCDs work? Oblong liquid crystals serve as light valves when stimulated with current from tiny transistors printed on an underlying thin film. The light is supplied by a backlight behind the screen. Most sets use **fluorescent backlighting**, though newer and hipper sets are using **LED backlighting**, which allows a wider color spectrum, deeper black level, more accurate whites, more brightness, and lower power consumption. LED-backlit sets are taking over the high end of the LCD market.

Toshiba

The Toshiba XZ900 Series set feature KIRA2 local-dimming LED backlighting. Whereas most LED-backlit sets use edge lighting, the KIRA2 sets have 512 controllable backlighting zones spread across the screen. This provides excellent brightness with greater uniformity than edge-lit sets.

There are two kinds of LED backlighting. The **local dimming** or **full array** type can vary pixels from bright to dim to off. This provides better black level, mitigating one of LCD's main drawbacks. Unfortunately there are fewer LEDs than pixels. That can cause light to spill over, causing halos around bright objects on dark backgrounds. In the drive to make LCD panels even slimmer, a new approach has become dominant: the **edge lit** type. It lines the screen's edges with LEDs. This allows for slimmer panels, but at the price of lower uniformity— brightness varies on the screen surface, which becomes apparent when the panel tries to reproduce a field of solid white or pale color. The experts prefer local dimming but it is getting increasingly hard to find.

The liquid crystals stand to attention when stimulated, then subside mechanically, and a bit slowly, leading to a lag that causes motion-related distortions called **motion artifacts**. The motion artifacts peculiar to LCD TVs are often called **image lag** or **motion smear**. When shopping for an LCD TV, look for the highest **refresh rate** you can find to minimize motion artifacts. Many new models have a **120 Hertz refresh rate**, which means they multiply the frame rate of incoming signals to 120 frames per second. Some go to a **240 Hertz** refresh rate. Such sets either double the standard 60-frame signal (LG, Toshiba, Vizio) or quadruple it (Samsung, Sony). This can look good, but only if the underlying circuitry keeps up. In public demos, this has been shown to reduce the blurring that occurs when the camera pans across something—say a face, or a newspaper. However, in demos not controlled by manufacturers, the difference is often much more subtle.

The increased refresh rate may not account for all of the difference. That's because many manufacturers combine it with **smoothing** or **anti-judder** circuitry, which is sometimes separately adjustable. This produces a more solid, though not necessarily more film-like, image.

One major problem with LCD sets is that most do not support a wide **viewing angle**. Sit off to the side and dark-color reproduction diminishes. The whole picture takes on a purplish hue. This problem affects some brands more than others—watch for it before you buy.

Dead pixels are a manufacturing defect to which LCDs are prone. They are likely to appear right out of the box. When you uncrate your set, look for them. If you see too many, take advantage of your warranty and exchange the set immediately. You're less likely to have this problem with top brands. This is one way in which high-end manufacturers like Samsung, Sharp, Sony, and Toshiba earn their price premiums.

Watch for these defects when buying an LCD TV. But don't let them put you off LCDs altogether. Some are better than others. And all LCDs share some major advantages. They are less prone to the **screen-door effect**—that is, you're less likely to notice the space between pixels. And though models vary, LCDs use less power than plasmas. Look for Energy Star certification when buying any TV. You can look up certified models at energystar.gov.

Flat-panel displays: plasma

Plasma panels come in sizes from 30 to 152 inches. Panasonic is the current record holder at the top end of the scale (got $150,000?).

Panasonic

The Panasonic TC-P65VT25 ($4300) is a 65-inch plasma with 3D-capable 1080p resolution. Like all plasmas, it is good at reproducing fast motion with minimal blurring. As a Viera Cast set, it also has a full panpoly of networked features including Skype video phone. Each set includes a pair of active-shutter 3D glasses.

7

Gas plasma technology produces light through tiny tubes filled with neon gas. When activated by electric current, the gas produces ultraviolet light, which in turn energizes the cell's colored phosphor coating. The underlying plate is a thick glass sandwich, making plasmas the heaviest flat panels—mounting a large one is a job for an experienced installer. Still, a huge display with no footprint is an attractive proposition in minimalist or space-starved homes.

Early-generation plasmas were not great performers. As noted above, the space between pixels tends to be more noticeable on a plasma than on an LCD, leading to the screen-door effect. However, the physical structure of the gas-filled vessels has been redesigned to improve plasma performance. Recent plasmas do a better job of containing and directing light, reducing bleeding and spillover, helped by a black stripe surrounding each cell. The depth of the cell enhances brightness, as well as color, because more colored phosphor can be packed into it. And the color filter is joined directly to the glass, with no air gap, reducing refraction of light and producing a purer image.

Most plasma displays have the full resolution necessary for HDTV. They offer some major performance advantages over competing LCD technology—wider viewing angle, less motion smear, and superior reproduction of deep black and shadow detail. Watch out for a blotchy effect known as **false contouring** in what should be smooth backgrounds. To reduce this effect, some plasma displays use an alternate video algorithm, a strategy called **error diffusion**.

You need to be aware of two other issues relating to plasma. One is **burn-in** or **image retention**. Like tube-based displays, a plasma panel can be damaged by static images that remain onscreen for long periods. Watch out for, among other things, video games and the bottom-screen crawl that accompanies some financial channels. To minimize the hazard, use moderate brightness and contrast settings (midpoint or lower) and control ambient light in the room. If burn-in does occur, run some full-motion, full-screen material for 24 hours. That may mitigate the damage.

Also, be warned that some plasmas behave oddly at **high elevations** where higher air pressure presses against the glass and causes a buzz. Some manufacturers adjust sets that are shipped to such areas.

Longevity was initially an issue—plasmas lose brightness as they age—but newer models are built to last. The 1080p models in Panasonic's 2008 Viera line take 100,000 hours to drop to half their original brightness. At 6.5 hours a day, that would take an average viewer 42 years. The 720p models take 60,000 hours or 25 years.

Finally, as you look at the many plasma brands in stores, be advised that only a handful of companies make plasmas—for themselves, and for other brandnames. The real powers in plasma manufacturing are Panasonic in Japan (Pioneer and Hitachi having withdrawn in 2009), LG and Samsung in Korea, and Chunghwa in China. Panasonic now makes Pioneer's panels, though Pioneer still does its own R&D, design, and underlying electronics.

Flat-panel displays: the future

OLED stands for **Organic Light Emitting Diode**. OLED displays are electroluminescent—in other words, they generate their own light. That provides a brighter and purer image than conventional LCD technology which relies on external backlighting. Like plasma displays, the current generation of OLED displays uses a heavy glass plate, but future generations are likely to use flexible polymers to provide bigger images, and less weight, at lower cost. Right now, OLEDs are available mainly in the tiny displays that animate certain models of cell phone and MP3 player. Sony began selling 11-inch OLED TVs in 2007, LG followed with a 15-inch version in 2009, followed by a 31-inch prototype in 2010, and the largest OLED display shown so far has been a 40-inch prototype from Samsung.

Rear-projection television

The term **rear projection** formerly referred only to a one-piece set that uses a trio of cathode ray tubes to project an image from behind. However, newer **microdisplay**-based sets substitute either DLP technology licensed from Texas Instruments or liquid crystal panels (see following sections). The term rear-projection is tricky—many front-projection TVs are able to throw an image from behind the screen. Mainstream RPTVs are the big, boxy sets you see in electronics stores. RPTV and flat-panel sizes overlap, and nowadays, most people prefer flat panels. As a result the rear-projection category is rapidly shrinking.

Rear-projection sets range in size from 40 to 70 inches, and occasionally up to 82, measured diagonally. Most models are in the fifties or sixties. In terms of dollars per square inch, they're the most economical HDTVs, and even the smallest projectors offer home-theater-worthy screen size. In fact, all else being equal, RPTVs with screen sizes in the

9

forties and fifties generally perform better than larger models. The smaller the set, the brighter the picture.

DLP projectors

DLP and other fixed-pixel technologies dominate both front- and rear-projection TV sales.

DLP was invented in 1987 by Dr. Larry Hornbeck of Texas Instruments. It produces an image by reflecting light off one or more chips containing thousands of tiny micro-mirrors. Texas Instruments used to refer to the underlying technology as **DMD (Digital Micro-mirror Device)**—now it's just called DLP technology. DLP projectors can be quite small, both because the chips themselves are smaller than the cathode ray tubes that drive old-style projectors, and because the chips reflect heat as well as light, so the need to dissipate heat is minimized. Their reflective technology wastes little light, enabling them to produce a huge, bright picture. They are not as good at producing true black as tube and plasma displays. However, the pixels blend better with DLP than with other fixed-pixel displays, making the picture more seamless.

TI has steadily refined DLP technology over successive generations of chips. The **xHD3** chip supports the 1080i HDTV format. However, though 1080i uses 1080 vertical pixels by 1920 horizontal pixels, the chip has just 960 horizontal pixels. Each one flips open and shut twice, simulating the effect of two pixels. The technique is sometimes called **wobbulation** (on Hewlett-Packard models) or **SmoothPicture**. Subsequent chips dropped the HD nomenclature and were simply labeled by resolution, either 720 by 1280 or 1080 by 1920 pixels. **DarkChip 4**—driven by cinema applications—provides a further 30 percent increase in contrast. The latest chip supports **3D video** for movies and games. 3D is now widely deployed in DLP-driven movie theaters and Mitsubishi offers 3D DLP rear-projectors for the home. Games are another killer app.

A DLP projector may use one chip or three. All other things being equal, **three-chip DLP** models are more expensive—but better suited for high-end home theater or public theaters, which require extremely high brightness. Their optical innards filter light into video's three primary colors (red, green, blue), routing each color to its own chip, then recombine the three light streams into one before the beam exits the lens at the front of the projector.

The more prevalent **one-chip DLP** models work differently, using a rotating **color wheel** to filter one color at a time, though it happens too quickly for the human eye to detect except in rare cases. Some eagle-eyed viewers claim to see a so-called **rainbow effect** that is most visible with bright objects rapidly moving against a dark background (such as a flashlight swinging in the dark). Newer models defeat the problem with much higher processing speed. Even on older models, it's usually so rare and fleeting that few people are bothered.

A color wheel must have at least three segments: red, green, and blue. That's a **three-part color wheel**. A **four-part color wheel** adds a colorless segment to increase brightness. In a **six-part color wheel**, the red, green, and blue segments are doubled, reducing the rainbow effect to (in my opinion) virtual nil. Mitsubishi's version of the six-part wheel uses red, green, and blue plus yellow, cyan, and magenta. This brings a huge increase in color fidelity—daffodils actually look yellow.

But the best color wheel may be no color wheel. DLP rear-projectors by Samsung use flashing **LEDs** (light-emitting diodes) to re-place the color wheel and the arc lamp that goes with it. These lamp-free projectors operate well even in rooms with lots of ambient light. Among the other benefits are no rainbow effect, longer life (20,000 hours), more uniform performance over the lifetime of the set, quicker turn-on, qui-eter operation, and as an environmental bonus, toxic mercury has been eliminated from the design. Unfortunately these LED-DLP sets per-ished with Samsung's withdrawal from rear-projection TV.

Another approach, by Mitsubishi, replaces the color wheel with **la-sers**. So far Mitsubishi's **LaserVue** sets are the only ones to do so. The big benefit is the elimination of the color wheel killing the rainbow ef-fect. The latest LaserVue line includes two models, 75 and 65 inches.

In terms of cost, DLP projectors are becoming increasingly practi-cal as their prices continue to drop. They can be purchased in the form of rear-projection sets (mostly from Mitsubishi) or front-projection sys-tems (from many brands). Be careful to distinguish between consumer-level models and pro models. The latter may lack features necessary for consumer video use, such as HD resolution, video processing, and the ability to switch between widescreen and non-widescreen programming.

LCD projectors

LCD is another microdisplay technology that competes head to head with DLP in both front- and rear-projectors. It is a particularly good value in high-definition front-projection displays.

Front- and rear-projection sets using a liquid crystal panel (or up to three panels) have existed for many years though production problems initially kept performance low and prices high. The first few generations of LCD projectors were plagued with an intrusively visible pixel (picture element) structure, poor color, and a high factory defect rate. However, more recent LCD projectors produce a bright picture with good color that stands up to daytime viewing.

There are two kinds of LCD-based projectors: the **reflective** type, which reflects light off a liquid crystal panel, and the older **transmissive** type, which sends light through the panel.

Reflective LCD technology is usually called **LCoS** (**liquid crystal on silicon**). Though slow out of the gate, it has matured to the point where some of the highest-performing projection sets use this technology, especially front-projectors from JVC and Sony.

The version of reflective LCD technology used by JVC is **D-ILA** (**direct-drive image light amplifier**). At the heart of D-ILA is a tiny 1.22-inch liquid crystal panel about the size of a DLP chip. Because D-ILA reflects 93 percent of the light coming from the projector's lamp, it is highly efficient, and produces a very bright picture.

Sony's version of reflective LCD technology is **SXRD** (**Silicon X-tal Reflective Display**). It uses three panels with what the manufacturers describes as a uniform liquid crystal cell gap, without any spacers in the image area, firmed up by an inorganic alignment layer. As a result the picture has a high pixel density while the space between pixels is as small as possible. And the process is robust enough for mass manufacturing at increasingly attractive prices.

D-ILA and SXRD are available in both front- and rear-projection versions and can offer full HD resolution of 1080 by 1920 pixels. The latest versions are simply stunning, with a deliciously fine-grained picture, plenty of brightness, and vivid color. If I were venturing into the market for front-projectors, I'd go for one of these.

Traditional CRT projectors

The traditional video display of choice for high-end videophiles is the **CRT**-based projection TV. CRT stands for **cathode ray tube** and a CRT projector uses three of those big, heavy tubes in video's primary colors of red, green, and blue to produce an image. The best CRT-based projectors can accurately display just about any video format that exists. Their performance in some areas, such as black level and shadow detail, remains unsurpassed. In the words of Joel Silver of the Imaging Science

Foundation, they are video's "gold standard."

CRT projectors make more demands on the consumer than newer display technologies. Good ones are costly. They are larger, heavier, and more difficult to mount than DLP or LCD projectors. The best CRT projectors are the largest ones with nine-inch (or larger) tubes. More affordable models with seven-inch tubes can also produce a beautiful picture but don't offer as much resolution. (A nine-inch CRT projector can display all the scanning lines in a 1080i HDTV picture; a seven-inch CRT projector usually cannot.) For best performance, CRT projectors require professional installation. They also need regular adjustment (called **convergence**) to keep all three tubes in alignment. Otherwise, color fringing will result, and will worsen over time. In addition, because green color phosphors age more rapidly than red and blue, a technician must occasionally adjust the **color balance**.

JVC

The JVC DLA-HD990 ($10,000) is an HD-capable front-projector based on reflective liquid crystal on silicon (LCoS) technology—specifically JVC's version, called D-ILA. It is THX certified for picture quality and ISF certified to ensure that a qualified techncian can optimize the picture.

13

Though CRT projectors have the potential to produce a fantastic picture, they do have a couple of performance disadvantages. One is **blooming**—when the tubes are overdriven, images get bent out of shape. Another is burn-in, or image retention—static images from videogames and other graphics-heavy material may leave a permanent imprint on the tubes. This is also a problem with plasma but it's a bigger one with tubes.

Just because the CRT isn't practical for everyone doesn't necessarily mean it's good for no one. Its position in the home theater market is analogous to the LP's position in the 1980s at the dawn of the CD era. At that time, the LP ceased to be the mainstream music-delivery format, and became instead a rallying point for audiophiles who declined to lower their standards to accommodate a developing digital technology. Like a Linn turntable, a nine-inch CRT projector is most suitable for the high-end consumer who's willing to go to extra trouble and expense to obtain the last possible increment of performance.

If you buy a used tube projector, make sure it has been **retubed**. Replacing the CRTs adds to the cost but ensures that you'll get the full benefit of your investment.

Projection screens

The most significant accessory for a front projector is a screen. Choosing the wrong screen can negate your investment in a good projector. The screen directly affects how you'll perceive the projector's performance (remember that when taking in projector demos!). Consider it part of the projection system's cost.

Gain, or the amount of light the screen reflects, is the key spec. Screens with a gain of 1.0 reflect back precisely as much light as they receive from the projector. Those with a gain of more than 1.0 offer a tradeoff. They provide more brightness to viewers sitting at the center but less to viewers sitting at the sides. Therefore high-gain screens provide more brightness, but less uniform brightness, a fact reflected in another specification, **viewing angle**.

Avoid extremely high-gain screens if you have a large family, or plan to host screening parties with many guests. After all, you want audience members at the sides to see a decent picture. A good limit would be a gain of 1.3—for instance, the Stewart Studiotek 130 (developed by the influential video expert Joe Kane) or the Da-Lite Cinema Vision. On the other hand, if your home theater is long and narrow, or you do most

14

of your viewing alone, the extra brightness afforded by a high-gain screen can do a lot to enhance the brightness of a projector. A high-gain screen also allows some projectors to work better in the presence of ambient room light—though reducing the light level in the room is a better solution for most than using a high-gain screen. High-gain screens are especially suitable for 3D due to the latter's reduction in brightness.

With fixed-pixel projectors on the ascendant, screens designed to compensate for their shortcomings have become a necessary accessory. These screens are literally grey, and that enables them to get a deeper black level and more shadow detail out of the projector. My home theater system features the Stewart FireHawk. The FireHawk's gain is slightly greater than 1.0 but some other DLP-friendly screens have lower gain. Example include Stewart's GrayHawk, with a gain of 0.95, and Da-Lite's High Contrast Da-Mat, with an even lower 0.8.

Though the following statement may attract derision from high-end videophiles, you may not need a screen at all if you're using a DLP or LCD projector. A clean wall surface painted a neutral white can serve as a stealth screen. This is not the best way to maximize the performance of a projector—but it may be an attractive option for someone who does not want to call attention to a home theater system when it is not in use. Another advantage of using a wall is that you can zoom the image in and out without having to make any binding decisions about how large it should be.

One of the advantages of using a screen is that it will have **black masking** around the edges. Masking helps frame the image and, by providing a contrasting element, makes it subjectively appear brighter and more vivid. Masking can be made adjustable to fit both widescreen and nonwidescreen **aspect ratios** (screen shapes).

Screens not mounted within frames can curl, and bend shapes on-screen. **Framed** or **tensioned** screens don't have this problem; they can sit on legs or their frames can be mounted to the wall. Don't want to see your screen when there's no show? Look at **roll-up** screens with motorized mechanisms that withdraw the screen when it's not needed.

Video processors

Because front-projectors produce such a huge picture, they mercilessly magnify the flaws in standard-definition and analog signals. Even if you are a digital video enthusiast, this is still an issue if you want to watch DVD, VHS tapes, laserdiscs, or other analog video sources.

To deal with this problem, some projectors function better with the help of **video processors** or **scalers**, which may be built into the projector (or a surround receiver or surround pre-pro) or bought as a separate device. These devices increase the number—and therefore reduce the size—of scan lines on the screen. Increasing the number of scan lines does not increase the amount of information in the picture, but does rearrange the existing picture information to better advantage, making the line structure less visually intrusive, and producing an image that more closely approximates the feel of film.

The crudest video processors are merely **line doublers**; **line quadruplers** or **line interpolators** do a more thorough job. Most consumer-level front- and rear-projection DTVs already have built-in video processors. The question, then, is whether an outboard processor would improve on the built-in one's performance. Going from a simple line doubler to a more sophisticated video processor can produce a visible difference. A high-end video processor used to be a five-figure investment. However, with a progressive-scan disc player and/or a video processor costing a few hundred dollars, you can get performance that would have cost tens of thousands just a few years ago.

To get the benefit of a video processor, your projector will need to generate the scan lines that produce the picture at or above a certain speed. It should have a **scanning rate** of at least 31.5kHz (kilohertz). That's the same scanning rate required by 1080i HDTV; 720p requires 45kHz. A line quadrupler requires a projector with a scanning rate of 63kHz and up. For 1080p you'll need 67.6kHz. However, there's more to mating projectors and processors than just keeping an eye on the scanning rate. A projector may have an optimum scanning rate, below its maximum rate, that maximizes light output and sharpness.

The increasing popularity of fixed-pixel displays that convert all signals to a single scan/refresh rate has led to a new category of video processor known as the **single-rate scaler**. Why invest in a sophisticated multi-format scaler when all you need is one that handles a single format and does it well? Some single-rate scalers are offered as accessories with matching projectors.

Warning: Don't go it alone! Speak to your projector's manufacturer, as well as your local high-end video dealer, about what kind of video processor (if any) is appropriate for your projector. Take your time and watch a few demonstrations—with *your* model of projector—and observe differences before and after processing.

16

DTV by the numbers

Videophile specsmanship is at its most intense in discussions of resolution. Quantifying the dots that make up the picture is both a science and a sport. It's also something you need to know before you commit to buying a DTV.

Understanding resolution

- resolution, native vs. measured
- resolution, vertical vs. horizontal
- lines of horizontal resolution

The spec known as **resolution** determines both **detail** (the amount of information in the picture) and **sharpness** (the amount of brightness or contrast at the edges of objects). It is the one of most important specs to look at when you're buying a TV and one of the most potentially misleading.

There are two kinds of resolution: native and measured. **Native resolution** indicates the maximum number of visible **pixels** (**picture elements**) that a video display is capable of delivering, as dictated by its physical structure and underlying electronics. **Measured resolution** indicates performance as measured on a test pattern. These two kinds of resolution may be interrelated but are not identical. Measured resolution may fall short of native resolution, but can never exceed it.

In digital television, native resolution is a straightforward pixel grid arranged in vertical columns and horizontal rows. For example, if a projector has **vertical resolution** of 720 pixels and **horizontal resolution** of 1280 pixels, on a spec sheet, that would read "720 x 1280" (or vice versa). Note that *the words vertical and horizontal refer to the method of counting, not the things being counted.* Vertical resolution describes horizontal rows of pixels, counted vertically, from the top of the screen to the bottom. Horizontal resolution describes vertical columns of pixels, counted horizontally, from side to side. (Because the pixels may not be placed precisely on top of one another, horizontal resolution may also be counted

17

as the number of pixels on each horizontal line.)

In analog television, resolution is measured a little differently—as **lines of horizontal resolution** or, more simply, **TV lines** (**TVL**). Do not confuse this spec with the number of horizontal scan lines. It refers to the number of visible vertical lines that can be counted horizontally—not across the full width of the screen, but across a horizontal distance no greater than the picture height. If you visualize it correctly, your mind will summon up a square made up of alternating black and white vertical lines.

How much resolution is enough? In the age of HDTV, you need at least 720 by 1280 pixels, and videophiles prefer 1080 by 1920. We'll discuss HDTV formats in the next section.

Blu-ray and DVD test discs allow you to test resolution and adjust picture settings on any set linked to a disc player. Note that DVDs are standard-, not high-definition. With those, you can measure lines of horizontal resolution only up to the original DVD format's limit of 540 TV lines (or less). However, new versions in the Blu-ray format can measure the full resolution of HDTV—for example, *Digital Video Essentials: HD Basics* (Joe Kane Productions). Take one of them to the store and run your own tests if you really want to see how TVs perform. (Don't be surprised if you get thrown out of the store.)

Meet HDTV, EDTV, and SDTV

Before we get into digital television, let's note what it replaced. The now-obsolete North American **analog television standard** is named **NTSC**, after the **National Television System Committee**, which finalized it in the early 1950s. It has a total of 525 scanning lines. That determines its vertical resolution, though the two numbers are not identical. That is because not all of the scan lines are **active** (meaning visible)—some are used for control purposes, to stabilize the picture, or to deliver closed captioning. Therefore the measurable and visible vertical resolution of an NTSC set is about 480 lines, with each horizontal line holding about 440 pixels.

Eighteen different formats go into the **digital television** (**DTV**) standard as defined by the **Advanced Television Systems Committee** (**ATSC**), the standard-setting body for which the DTV standard is officially named. However, they fall neatly into three groups. **High-definition television** (**HDTV**) offers a picture with roughly twice as much information as analog TV. **Enhanced-definition television**

(**EDTV**) offers a picture just slightly sharper and more detailed, but much cleaner, than analog, thanks to its use of progressive scanning (see below). And finally, **standard-definition television** (**SDTV**) is a lower-resolution format offering modest advantages over analog television. Note that the SDTV designation once referred to the EDTV format—older product literature may refer to EDTV as SDTV. And the fun has just begun!

For the sharpest picture, join the community of videophiles who want HDTV to improve the quality of life. HDTV sets have the sharpest pictures, making them the best choice for big-screen viewing. EDTV and SDTV have more relevance as transmission formats than as hardware formats. However, some DLP and flat-panel displays do have EDTV resolution, give or take a few pixels, and do a good job of displaying DVD and downconverted HDTV.

Following are two tables that'll help you better understand digital television technology. The first one sums up all 18 of the video compression formats sanctioned by the ATSC. These are the formats in which DTV may be transmitted over the air (satellite and cable transmissions don't always conform to these numbers). The second table describes the formats used in some DTV sets.

Vertical lines	Pixels	Aspect ratio	Picture rate
1080	1920	16:9	60i, 30p, 24p
720	1280	16:9	60p, 30p, 24p
480	704	16:9 or 4:3	60p, 60i, 30p, 24p
480	640	4:3	60p, 60i, 30p, 24p

The first column above represents the number of horizontal scanning lines that make up the picture and determine its **vertical resolution**. In the second column is the number of pixels per scanning line, which determines **horizontal resolution**. **Aspect ratio** is screen shape (16:9 is widescreen, 4:3 a conventional screen). As you glance down the far righthand column, you'll notice that the number of pictures transmitted per second is variable, and that pictures may be transmitted as full **frames** (**p** is for progressive scanning) or as half-frames, known as **fields** (**i** is for interlaced scanning). The distinction between progressive and interlaced scanning is discussed further a few pages down.

Format	Resolution	Aspect ratio	Picture refresh rate
1080p HDTV	1080 x 1920	16:9	60 frames per second
1080i HDTV	1080 x 1920	16:9 or 4:3	60 fields per second
720p HDTV	720 x 1280	16:9	60 frames per second
480p EDTV	480 x 848	16:9	60 frames per second
480i SDTV	480 x 640	4:3	60 fields per second

This second table will interest the DTV shopper more—these formats are not for transmission, but for actual display on a DTV you may be considering. Some are direct implementations of ATSC formats; others are conversions to a display's native format.

At top is the much-talked-about 1080p format, also known under the marketing term **Full HD**. It does not exist in the ATSC specs—in other words, there is no 1080p broadcast programming—but many new HDTVs have 1080p signal processing. This is mainly an upconversion format. When the set receives any kind of signal, it converts to 1080p. This may provide a more seamless picture, if done well, but doesn't really afford any more real picture information. The first *serious* use of 1080p is in Blu-ray. Along with the ill-fated HD DVD, it's the first signal source available to consumers that supports true 1080p.

Next is the 1080i format used in some HDTV broadcasts. It provides the maximum number of scan lines in an interlaced format. (See the next section on "Interlaced vs. progressive scanning.")

The last three items describe lower-priced displays, high-definition and enhanced-definition. One of them provides the same resolution as the 720p ATSC format with progressive scanning. Others provide the same 480 vertical pixels as the ATSC's EDTV format but stretch the horizontal pixels out to 848 to optimize the imaging chip for widescreen use. At the bottom is 480i SDTV, the closest digital equivalent to the now-gone analog broadcast standard.

At the far right, the numbers have all suddenly standardized at 60 fields or frames per second. That's to minimize flicker, which would occur at lower **screen refresh** rates. Note, however, that the formats delivering 60 full frames per second have a major advantage over those delivering only 60 fields (half-frames) per second.

These are just some of the ways in which DTV may be implemented. You'll sometimes see pixel-grid numbers other than these.

Interlaced vs. progressive scanning

When the video salesperson mentions 1080p, what do those numbers and letters mean? The numbers are easy—they're the scan lines and pixels that produce the picture and determine its vertical resolution. The letters, on the other hand, refer to interlaced and progressive scanning, the two different methods that analog TVs and computer monitors use to produce a picture. A DTV set may use either method.

Interlaced scanning uses two alternating **fields** of horizontal scan lines to form a **frame**—that is, a full still picture. First, one field worth of lines traces a ghostly half-picture from top to bottom, with spaces in between each line. Then a second field comes along and fills in the gaps, interlacing with the first.

Progressive scanning is simpler. It is not measured in fields—only in frames, as the horizontal scan lines trace a full frame from top to bottom. Thus is created, frame by frame, the illusion of moving pictures. This is how films have been made from Sir Charles Chaplin's two-reelers to the latest blockbuster. Most DTVs now use progressive scanning.

If you own an analog TV and a PC, then you already own one screen of each type. Your old analog set uses interlacing, while your computer monitor is progressive. Both formats found their way into the DTV standard as a compromise between the TV makers (some of which favored interlacing) and the computer industry (which favored progressive).

Interlaced scanning makes it possible for DTV makers to deliver lots of scan lines while still keeping down video bandwidth and costs. It's twice as efficient because a field contains only half as much information as a frame. However, few HDTVs still use 1080i with interlaced scanning. Most now use 1080p or 720p with progressive scanning. 1080i is mainly a broadcast format.

1080i vs. 720p

HDTV comes in two forms. Neither is automatically superior—each one does something better than the other. Both are considered to be HDTV by the Advanced Television Systems Committee (ATSC), which designated the formats following years of research and debate, and by the Consumer Electronics Association, an industry trade group.

The **1080i** format produces a picture with 1080 horizontal scan lines, in two halves, interlaced together (as explained above). The **720p** format produces a picture with 720 lines, one full frame at a time, with progressive scanning—no interlacing.

The key distinction between 1080i and 720p is spatial versus temporal resolution. Better **spatial resolution** is the strength of 1080i. Because it uses more scan lines per frame, it produces a sharper and more detailed picture when the image is frozen or barely moving.

Stronger **temporal resolution** is the advantage of 720p. It can reproduce rapidly moving objects with less blurring, as well as cleaner graphics, because it uses progressive scanning that leaves the frame intact. Most tube and projection displays can't display all the lines in the 1080i format anyway. The scan lines blur together, though the picture may still be dazzling compared to an analog set.

The 1080p format, delivered by Blu-ray, combines the strengths of 1080i and 720p—as we'll see in the next section.

Except for PBS, whose affiliates are free to use either 1080i or 720p, the major broadcasters have endorsed one format or the other. CBS and NBC are firmly in the 1080i camp. ABC, after initially rebelling against the whole notion of HDTV, has become a stalwart supporter of 720p. Rupert Murdoch's Fox Network was initially the least enthusiastic about HDTV in general, electing to deliver widescreen 480p instead— still a big improvement over analog—but eventually went 720p, or 1080i in some of its cable channels. That doesn't mean you have to buy a separate DTV for each format—any DTV set you buy should display pictures in any format by upconverting or downconverting to its native resolution (see next section).

1080p: HDTV at its best

HDTV in a **1080p** format, also known as **Full HD**, has moved from pro/archival use to consumer use. This format combines the strengths of 1080i and 720p, producing a 1080-line picture in full frames.

Many HDTVs already support 1080p as an upconversion format— in other words, they display other signal formats in 1080p to gain a small edge in performance. However, true 1080p *signals* were limited to industrial users until recently. Among other problems, they are too large to fit onto a DVD or through a broadcast channel, and therefore were offered only as a display format for upconverted signals.

The advent of the Blu-ray disc format has brought 1080p to consumers for the first time. You probably won't get it off the air any time soon but at least you can get it (in order of quality) via disc, telco-fiber, cable, or satellite.

Spec-wise, there's no doubt that 1080p is the state of the art in HDTV. However, there are limits to what the human eye can see. Even if your vision is perfect, you can't see the pixels on a 720p display beyond a certain distance—so whether you really need 1080p is a function of viewing distance and screen size. For smaller HDTVs at correct viewing distances, 1080p is overkill (though it certainly doesn't hurt).

24fps: the movie lover's frame rate

In addition to 1080p, another bit of alphanumeric magic is creeping onto spec sheets. When you see **24fps**, that indicates that movie-based material can pass from some high-def signal sources (like Blu-ray) to a compatible set at 24 frames per second. This is the same frame rate movie projectors use. It's rather flickery, though, and that's why movie theaters double the frame rate from 24 to 48. In television, the frame rate is converted from 24 to either 30 or 60, a process called 3/2 pulldown (see below). How well your HDTV, disc player, or other signal source performs 3/2 pulldown matters a lot more than whether the source is 24fps. Note that 24fps and 3/2 pulldown apply only to film-based material, not programming shot on videotape.

Upconversion and downconversion

A key feature of any digital set, especially an HDTV set, is its ability to convert video from other formats to the **native resolution** that best fits the set itself. Every display is physically and electronically optimized to work best with a certain number of pixels or scan lines, which determines detail and sharpness. For instance, many of the most desirable new HDTV models have a native vertical resolution of 1080p, meaning they display a picture with 1080 scan lines in one pass per frame.

What happens when a 1080p set gets something other than a 1080p signal? After all, there's no such thing as 1080p off the air—HDTV broadcasts are either 1080i or 720p—and because the amount of high-def programming available is limited, some channels or programs may operate at lower resolution. When a native 1080p set receives, say, a 480p signal from a DVD player, or an analog channel from cable, it may **upconvert** the signal to 1080p.

That doesn't mean it turns fuzzy signals into sharp ones. It just means the set multiplies the 480 lines to 1080. There may be 1080 scanning lines onscreen but they display 480 lines worth of information. This is analogous to what happens when you play a soft-sounding audiocassette through a high-end stereo system. It still feels like a soft-sounding audiocassette.

By the same token, an EDTV set has to **downconvert** higher-resolution 1080i signals to its own native 480p resolution, which is like putting a CD in a boombox. The boombox will still sound like a boombox though the CD will allow it to sound as good as a boombox can sound.

DTV sets will up- or downconvert in different ways. For instance, some native 1080i rear-projectors will upconvert 480p and analog signals to 1080i, while others will upconvert them to 540p (note that 540 times 2 equals 1080—the multiple makes for a clean conversion) or just display them at the original 480p.

Note that sets that convert film-based programming from 480i to other native resolutions include **3/2 pulldown** to compensate for the mayhem that occurs when 24-frames-per-second film is converted to 30-frames-per-second video. As an alternative, 3/2 pulldown can also be performed by a progressive-scan disc player. For more information about this topic see "Disc Players/Progressive-scan DVD-Video players."

Conversion determines how your DTV will perform with different sources of programming and you should consider every possible angle.

Interpreting TV specs

- video S/N
- chroma AM S/N
- chroma PM S/N
- brightness and lumens
- color temperature
- color gamut

Resolution is not the only way of quantifying a display's performance. Here are some more things you're likely to see on TV spec sheets. Note that this discussion is about specifications, not TV controls. The controls don't dictate the specifications. They merely adjust the set's performance within the limits of its specs.

Expressed in decibels (dB), **video S/N (signal-to-noise) ratio** is another key spec. It describes the amount of noise, snow, or other visual contamination in the brightness portion of the video signal. The **chroma S/N ratios** do the same for the color portion, which is separate. There are actually two kinds of chroma S/N. **Chroma AM** (amplitude modulation) describes how much picture noise affects saturation or intensity of color. **Chroma PM** (phase modulation) describes how much noise affects hue, or color fidelity (a fire engine should not be purple). If you want a clean picture, pay attention to these less often quoted specs.

Brightness—again, the spec, not the picture control—is measured in the whitest part of the picture with the contrast control at its maximum setting. (And by the way, only a fool would actually watch a movie that way.) The spec may be given in two ways. A projector's light output is given in **lumens** or **ANSI lumens** (per the American National Standards Institute). ANSI lumens are measured differently than regular lumens. Using the ANSI specification method, numbers tend to be dramatically lower. Reflected light—what you actually see on the screen—is measured in **footlamberts (ftL)**.

Although some direct-view tube TVs can achieve more than 100 ftL, the broadcast industry standard spec for professional direct-view video monitors is about 30 ftL. Spec inflation is rampant here simply because what is feasible for measuring is nowhere near appropriate for normal viewing. A CRT-based direct-view or projection TV should *never* be viewed at maximum light output. In consumer-level tube sets, even less than 30 ftL will stress the picture tube, causing a problem called **blooming**, visible as increased **geometric distortion** (objects bent out of shape) and burning both the tube and the dotted chemical color **phosphors** within the tube. In non-CRT sets, excessive brightness wears out the lamp or backlight more quickly than normal brightness. With projectors in general, less than half of the 30 ftL figure is normal, because they are designed to be used in darkened rooms.

Don't get hung up on these numbers. Instead, as you eyeball projectors in showrooms, try to get the picture projected to the same size that your own screen will be (brightness drops as picture size increases). And ask the dealer to adjust the light level to approximate what's going on at home. Keep in mind that projection lamps lose light output as they age, so neither the specs nor the showroom demo will reveal everything about the longterm experience of using the projector in your home.

Many videophiles think the correct **color temperature** (or color balance) is indispensable to achieving a truly filmlike image. Measured in **Kelvins** (some prefer to say "degrees Kelvin"), color temp should be precisely 6500 according to the television standard. That's neutral white, and therefore a good place to start if you want accurate color. Unfortunately, most sets sold today are not shipped in a mode that meets the standard because manufacturers think turning the whites blue, and thus cranking up brightness levels, will make their TVs more appealing under bright showroom lights. Over time the effect looks more and more artificial as the viewer realizes that all other colors have been skewed to generate a bluer, brighter white. Most sets offer the option of choosing 6500 Kelvins from among several settings (often called the **movie mode** or **film mode**). Many others can be set to the standard by a technician, such as those certified by the Imaging Science Foundation (visit imagingscience.com for more details). Finally, note that color temp may not be fully **linear**, instead varying according to the level of brightness. But setting it as close to the standard as possible will bring an enormous improvement in color fidelity.

Color gamut, the range of colors reproduced, has become the focus of fad features on DTV spec sheets. These include **Deep Color** and **x.v. Color**, also known as **xvYCC**. While these expanded color gamuts are being built into some new sets, they will affect the picture only if they are accommodated in production gear and source material.

MPEG-4 AVC and the future of HDTV

Video compression is a big part of digital television. Without it, the stream of zeroes and ones that carries DTV would be too fat to enter our homes via antenna, satellite, cable, and broadband connections. The latest form of DTV compression, **MPEG-4 AVC**, performs new feats of compression that will affect both the quality of digital television and the ways it can travel. Its most immediate effect is seen in satellite TV and some Blu-ray releases.

Right now the dominant form of DTV compression is **MPEG-2**. MPEG is the Moving Picture Experts Group, a Geneva, Switzerland based outfit that brings together standards experts from all over the world. Another set of MPEG video-compression formats, known as MPEG-4, has been around for awhile and has been used mainly to deliver low-resolution video in PC and cell phone applications. MPEG-4 AVC (**Advanced Video Coding**) is a higher-quality variation that does more with fewer bits than any previous MPEG video format. Its main competitor is **VC-1**, another MPEG-4 variant that originally surfaced in Microsoft's Windows Media Player 9.

AVC, VC-1, and the outgoing MPEG-2 are all included in the specs for Blu-ray, which can provide superb next-generation HDTV. However, AVC has also been tapped for use in DirecTV's HD local-channel lineup, followed by the Dish Network. The satellite providers like AVC because it enables them to deliver even more channels than their newly launched satellites would otherwise allow. MPEG-4 AVC is therefore the future of HDTV.

The shape of DTV

Television has changed its shape. You hear a lot about the transition from analog to digital TV and the potential wonders of HDTV. Just as important, but less heralded, is the transition from old-style narrow television to the wider screens of today.

Unfortunately, your local retailer may not yet be widescreen-hip. Visit a chain store selling HDTVs and you're likely to see a shocking display of technical illiteracy: the stretching of nonwidescreen signals on widescreen displays. There is nothing attractive about a video image that is stretched out of shape. Please, don't try this at home!

Screen shape, 16:9 vs. 4:3

If you buy a DTV, you'll have two choices of screen shape, also known as **aspect ratio**. Widescreen sets have a **16:9** ratio of screen width to screen height, similar to that used by most contemporary filmmakers and HDTV programs. Other sets mimic the **4:3** ratio of older movies and analog television. (You may read those numbers as "sixteen by nine" and "four by three.") To reduce those numbers to common denominators, that's 16:9 vs. 12:9, or 1.78:1 vs. 1.33:1, so widescreen proportions are about a third wider.

Until the 1950s, all movies and all television programs were made in a ratio of 4:3. Films were first to break the mold. Mid-20th-century movie moguls thought widescreen film would be a good stick with which to beat television. They figured audiences would flock to see films made in a wider aspect ratio.

With such films as *Ben Hur* (1959) and *Spartacus* (1960), both of which won Oscars for their (literally) spectacular cinematography, Hollywood redefined action movies and ushered in the age of widescreen viewing. These films were made in CinemaScope, which boasted a super-wide aspect ratio of 2.35:1.

Widescreen moviemaking didn't adhere to a strict standard. Some filmmakers preferred a less extreme width. Some even designed films to be viewed in any of several aspect ratios. A good example would be Alfred Hitchcock's *North by Northwest* (1959), which the Master of Suspense intended to be viewed at 2:1, 1.33:1, or anything in between.

Today, most filmmakers favor an aspect ratio of 1.85:1, though some prefer a wider 2.40:1. The former is close to the 1.78:1 (16:9) aspect ratio adopted for widescreen formats in the DTV standard. HD television programming matches the standard precisely.

Panning-and-scanning

Typically, it is the width—not the height—that is trimmed when a widescreen movie has to fit onto a 4:3 TV screen. The traditional means adopted for this cinematic surgery is **panning-and-scanning**, which allows a technician to choose which side of the screen should be cropped the most. This is the technique used to transfer most films to video for VHS and other old-style analog media. Panned-and-scanned DVD releases are labeled **full screen** (as opposed to widescreen) though they don't display the full image.

Ever watch a scene with two people talking face to face, but the backs of their heads are cut off, with a veritable sea of space between their truncated faces? Or maybe one is cut out entirely, appearing to yap at the side of the frame? Or maybe the camera seems to be weaving drunkenly to and fro between them? Ever see part of a movie that just doesn't make sense—like a character reacting to something outside the frame that you can't see? Do camera moves in general appear clumsy, as though the camera operator were having a really bad day?

Those are all the effects of butchering widescreen films with the pan-and-scan technique. Today, most directors and cinematographers work defensively, visualizing a 4:3 sweet spot as they compose the widescreen image, and making sure nothing significant falls outside it. This limits their use of the widescreen medium. It also means that older widescreen movies, made before the age of pan-and-scan savviness, often look terrible when trimmed.

However, panning-and-scanning is a rational response to the fact that people watching 4:3 TVs like their rented movies to fill the whole screen, without the black bars used in letterboxing. Another advantage of panning-and-scanning is that filling the screen helps avoid uneven burn-in of plasma and tube-based displays.

To accommodate both videophiles and average movie lovers, some two-sided DVD movie releases have a widescreen version on one side and a panned-and-scanned version on the other.

Letterboxing

As an alternative to panning-and-scanning, the video industry of the 1980s came up with **letterboxing**. This technique fits a widescreen image onto a narrower screen by placing blank (black or grey) bars at the top and bottom of the screen, thus allowing the full width to be viewed at the expense of height. Letterboxing was a popular stylistic gimmick among music-video directors of the '80s.

Letterboxing has drawbacks, reducing picture height and resolution. It fails to take advantage of a screen's total height. It also reduces resolution because fewer of the scan lines that produce the picture are devoted to the image itself. Some are used to record the blank letterbox bars, which sort of turns the TV into a six-shooter loaded with four bullets and two blanks. Most problematic of all, letterboxing causes uneven burn-in of plasma panels and the cathode ray tubes in direct-view and CRT-based rear-projection TVs (though the problem is eliminated in non-CRT displays, such as LCD and DLP).

These problems haven't prevented letterboxing from proliferating in several formats. Letterboxed laserdiscs used to be quite common. For many years they were the format of choice for videophiles. Even letterboxed DVD and VHS releases are not unknown, though further reducing the already poor resolution of VHS is not necessarily a good idea (don't even try reading the end credits). Blu-ray and DVD use letterboxing on the sides for pre-widescreen 4:3 movies, and on the top/bottom for contemporary movies wider than 16:9.

On a widescreen DTV fed a letterboxed signal, the blank bars appear not only at top and bottom, but on all sides. Selecting the set's letterboxing (or zoom) mode will enable the image to fill the screen but will not counteract the reduced resolution inherent in letterboxing.

Anamorphic: a better mousetrap

Fortunately, there's an alternative to letterboxing. It's called **ana-**

morphic video, and it's become the norm in the majority of widescreen DVD movie releases. (Blu-ray, like HDTV widescreen formats in general, is not anamorphic.) Anamorphic DVDs may be labeled in any of several ways, including:

- anamorphic widescreen
- enhanced for widescreen TVs
- enhanced for 16:9 TVs

Or they may not be labeled at all. Even so, if you have a widescreen DTV set, seeking out this kind of software would be worth your while. You'll need a DTV set with an anamorphic mode to get the benefit of anamorphic DVD movie releases. All consumer-level DTVs support anamorphic video nowadays.

How does anamorphic video work? To oversimplify a little, it squeezes and unsqueezes the picture to achieve maximum sharpness. Maybe that suggests the picture is undergoing some kind of degradation, but when you look at the alternatives—the truncation of the pan-and-scan technique, the bulky black or grey framing of letterboxing—anamorphic looks like the best option. It more or less fills the screen and the picture looks as sharp as the delivery medium will allow.

Actually, "squeezing" isn't the best way to describe the anamorphic process. Technically, what's happening is that an anamorphic DVD devotes more scan lines to the image and fewer to the blank bars framing the screen. Then the video display, correctly set to display an anamorphic picture, and playing anamorphic software, will reproduce a fullscreen picture with greater resolution. In the process, it also corrects the misshapen image that would appear without anamorphic decoding.

Many sets recognize the anamorphic mode and adjust to it automatically. However, beware of sets that automatically lock into 16:9 mode when receiving 480p progressive DVD input. They may incorrectly apply the anamorphic process to non-anamorphic material (for example, an older movie such as *Casablanca*).

The widescreen dilemma

Is widescreen video an inevitable form of progress, a major step forward? For the most part, yes, but it comes at a cost.

Entering the widescreen world, by buying a widescreen set, involves a tradeoff. Yes, you'll be able to watch 16:9 anamorphic DVDs and widescreen Blu-rays in all their glory. But 4:3 material, including many favorite older movies and current TV shows, don't occupy the full width of a 16:9 set. You'll have to accept either blank bars at left and right or (the greater of two evils, in my opinion) horizontal stretching of the image. Disconcertingly stretched images are all too common in stores—retailers really should know better.

Simply put, nonwidescreen material looks better on a nonwidescreen set, while widescreen material looks better on a widescreen set. As a practical matter, if you own a 4:3 set, you'll want to learn how to set your DVD player in the **4:3 letterbox** mode, which will correctly fit an anamorphic 16:9 image to the 4:3 screen by eliminating a few scanning lines. If you own a 16:9 set, Blu-ray is the best option, anamorphic DVD is second best, followed by the letterboxed version, though it won't have as much resolution. Panned-and-scanned or full-screen video transfers—especially on low-resolution VHS—are lowest in the videophile pecking order.

Aspect ratio control

Your DTV may come with several aspect-ratio modes to tailor various programming shapes to its screen shape. It may place blank bands at the top/bottom or left/right. It may blow up the image to fill the screen, cropping out some material. It may even do the latter in combination with some deliberate bending of shapes at the sides. You can choose the mode that suits the program material and your own tastes.

If you buy a 16:9 set, you'll want one that can correctly display both anamorphic and nonanamorphic DVD. Note that some sets will automatically go into anamorphic mode when receiving a 480p (480-line progressive) input. If the movie is not widescreen, it will suffer from the anamorphic squeeze, with severe bending of shapes.

3DTV

As flat panel TV prices have plummeted, manufacturers have desperately sought new ways to sell high-end sets. 3D is one of the technologies on which they're pinning their hopes.

History of 3D in photography and movies

A 3D image, in general, creates an illusion of greater depth perception by presenting each eye with slightly different images. This is known as **stereoscopy**. It goes back to the 19th century, when it was first used for still images. 3D for motion pictures made its debut in 1922 with *The Power of Love*. A major wave of 3D films arrived in the 1950s including *Bwana Devil* (1952), *House of Wax* (1953), and Alfred Hitchcock's *Dial M for Murder* (1954). 3D has continued to arrive in waves, with another major one in the 1990s. Of course the current cinematic and video revival of 3D has been led by *Avatar* (2010).

Basic types of 3DTV

What kind of TV is the best match for 3D processes? The leading contender is plasma, due to its inherently high motion performance. LED-backlit LCD sets can also produce solid 3D effects. Another alternative is DLP. For 3D front-projection, use a higher-gain screen to offset the reduction in brightness.

Because they require high-end video processing, 3DTVs also work well in 2D mode. Some provide **2D-to-3D conversion**, though this does not provide the optimum 3D experience.

There are many types of 3D technology. One way to sort them out is by looking at the glasses required to decode images. The most common type of 3D in its current video revival is **frame sequential**, which displays alternating frames for a viewer's left and right eyes. Battery-powered **active-shutter glasses** are synchronized by an infrared transmitter on the 3DTV and a receiver in the glasses. Other types of 3D include **anaglyphic**, which uses glasses with different-colored lenses, red and cyan; **polarization**, which uses passive polarized glasses; and **autostereoscopy**, which requires no glasses.

3DTV technology is baked into the panel, so whatever TV you buy will dictate the kind of 3D available to you. For instance, Panasonic and Samsung use frame sequential technology with active-shutter glasses—and each has its own proprietary version. Vizio uses polarization with passive glasses that look like sunglasses.

3DTV formats via HDMI

The HDMI interface is the preferred method for getting 3D images

MF

3D formats may use either the top-and-bottom method or the side-by-side method. In each case, there is a reduction in brightness, and depending on the form of transmission, there may be a reduction in resolution as well.

from 3D source components to 3DTVs. The first version of HDMI to handle 3D was 1.4, quickly succeeded by 1.4a.

HDMI 1.4 supports a variety of 3D formats including frame packing, checkerboard, interlaced, line-by-line, alternate frame, and 2D+depth. For 3D video delivered to the home, the two most significant formats are: **top-and-bottom**, and **side-by-side**. Visualize a frame split in half, either horizontally or vertically, and you'll get the idea. The 3DTV processors separate and scale this signal into individual left and right images. The active-shutter glasses direct each image to the appropriate eye, resulting in a single 3D image.

Blu-ray, with its capacious bit bucket, uses **frame packing**, the best possible form of the top-and-bottom method, with full 1080p resolution (1080 by 1920 pixels) in each eye.

DTV stations broadcasting top-and-bottom 3D cut vertical resolution in half. So 1080i video, normally 1080 by 1920 pixels, goes from 1080 vertical pixels to 540; 720p video, normally 720 by 1280 pixels, goes from 720 vertical pixels to 360, a serious loss of resolution.

Bandwidth-starved media like satellite and cable, which are rolling out their 3D services in 2010-2011, use side-by-side 3D and cut horizontal resolution in half. So 1080i video, normally 1080 by 1920 pixels, goes from 1920 horizontal pixels to 960. At present there is not a 720p side-by-side format in HDMI 1.4a.

HDMI 1.4a adds a few more 3D formats including frame-packing top-and-bottom at 2205 by 1920 pixels, which supports full 1080p resolution. For video games there's a full 720p version. To ensure maximum 3D compatibility, make sure any TV, Blu-ray player, or source component you buy meets the HDMI 1.4a specs.

The lemmings approach the cliff

I wish I could enthusiastically recommend 3DTV to my readers but I have several reasons not to. One is a relative lack of industry-wide standards—each 3DTV maker seems to have its own approach, so you can't be sure one's glasses will work with another's 3DTV. Multiple standards groups are now at work, which may help someday, though not as much as a *single* standards group.

Another problem with 3D, at least for a minority of viewers, is that it simply doesn't work. Not everyone's eyes and brain mesh with the various 3D technologies.

There are health concerns. A 3D warning given on Samsung's Australian site blurts out a whole laundry list of complications including epileptic seizure, stroke, altered vision, lightheadedness, dizziness, eye or muscle twitching, confusion, nausea, loss of awareness, convulsions, cramps, disorientation, motion sickness, perceptual aftereffects, eyestrain, decreased postural stability, damage to eyesight, headaches, and fatigue. While these warnings may be overblown, parents should limit the number of hours kids are exposed to 3D. If watching 3D bothers you, stop watching it. People with serious eye- or brain-related health problems may wish to avoid it altogether.

But the most persuasive argument against 3D is aesthetic. It comes from Pulitzer-winning film critic Roger Ebert, who writes: "When you see Lawrence of Arabia growing from a speck as he rides toward you across the desert, are you thinking, 'Look how slowly he grows against the horizon' or 'I wish this were 3D'?" Perhaps 3D's effect is not so much depth enhancement as depth exaggeration. Time-honored cinematic techniques, such as the use of perspective, may have the artistic high ground.

Home theater has been greatly improved by technologies introduced over the past decade or two—including HDTV, flat panels, and lossless surround—but 3DTV is not one of them. It is the most overrated new (actually not so new) video technology of recent years. And its implementation has been scattershot at best, with dueling screen technologies and numerous signal formats.

I fully understand that some readers may not share my skepticism. But unless you've fallen in love with 3D at the cineplex, feel free to give the home version a miss.

Tuners and other features

Friends, the world is changing. We have just seen a major transformation—the transition from analog to digital broadcasting. Analog broadcasting ceased on June 12, 2009 by an act of Congress.

Owners of analog TVs needn't fret about the transition to DTV. They can still use their old sets with set-top boxes. As for cable, the Federal Communications Commission passed "must carry" rules mandating that cable systems carry all digital over-the-air channels and make them available to all sets, digital or analog—either converting the signal at the head end or providing convertors to subscribers.

Digital television signals reach every major market in the U.S. To check the availability of DTV broadcasts in your area, consult the dtv.gov or tvfool.com websites. DTV is also available in an increasing number of cable and satellite channels. Over the last couple of years, cable, satellite, and telco TV providers have steadily escalated their HD offerings, each one fiercely insisting it offers the most HD options. This is nothing if not healthy.

Will you have a digital tuner with that?

To receive over-the-air signals, a video display needs a tuner. (Satellite, cable, and telco delivery require set-top boxes.) A **digital tuner** (or **ATSC tuner**) is one that receives digital channels. The **analog tuner** (**NTSC tuner**) is obsolete. Formerly, sets with digital tuners were called **integrated DTVs** while those without digital tuners were designated **DTV ready**. Now any set called an **HDTV**, an **EDTV**, or an **SDTV** is presumed to be an integrated set with digital tuner built into it. Any set that can display DTV but requires a separate tuner is classified as a **monitor**. An outboard channel selector is still called a **tuner**.

The distinction between a tuner-equipped DTV and a tuner-less monitor is a useful one. Some older DTVs lack digital tuners (though a government mandate is changing that for newer ones). Originally, this was because the specs governing the digital tuner were still in flux, as interactive features and copy-prevention systems evolved. For DTV buyers whose video feed comes from satellite or cable or telco, a monitor (without built-in digital tuner) still makes sense. An outboard tuner can always be added later if you change your mind.

Except in front-projection, DTV tuners are a legal requirement. The Federal Communications Commission voted in August 2002 to require that DTVs come with digital broadcast tuners—not all at once, but gradually. Fifty percent of DTVs 36 inches or larger were required to have digital tuners by July 1, 2004, with the rest of that size class following a year later. The phase-in for DTVs from 25 to 36 inches occurred between July 1, 2005 and March 1, 2006. By December 31, 2006 all digital TVs, recorders, and other devices with analog tuners were required to have digital tuners as well. A few rogue manufacturers, trying to cut costs in a cutthroat market, flouted the federal rules but were fined for breaking the law. Critics point to the fact that most Americans don't make use of over-the-air reception, depending instead on satellite or cable, and assert that adding the cost of a digital tuner to every DTV is unfair. Others defend the move, saying that it will speed the penetration of DTV, and that even people who depend on satellite or cable service still use antenna reception in the bedroom or other parts of the home.

Digital cable readiness

The latest news in digital cable readiness is that *it's the law*! You can thank the FCC for this. Finally acting on a 1992 federal law, the FCC began enforcing an **encryption ban** in July 2007. What this means is that the security function once exclusively provided by cable-provided boxes can now be provided by a card-enabled cable-ready TV, by a card-enabled cable box rented from the cable operator, or by a storebought box—giving consumers wide latitude in how to handle their cable-TV connections. Your cable operator would prefer that you rent a box but can no longer compel you to do so. The following backgrounder will explain how we got to this point and where we might go in the future.

The term **cable-ready** has long been a battleground for government, television manufacturers, and the cable industry. Consumers need a simple way to connect a cable TV feed to a DTV and receive digital programming without having to go through a cable-company-supplied set-top box. Such a cable TV **plug and play** standard now exists thanks to an agreement signed by cable operators and TV makers on December 12, 2002. It uses a Motorola-developed scrambling method known as **DFAST** (Dynamic Feedback Arrangement Scrambling Technique). Plug-and-play DTVs and other products are designated **digital cable ready** or **DCR**. The more technical term is **unidirectional digital cable**. The standard was supposed to have been enabled by July 1, 2004 in all cable systems with an activated channel capacity of 750MHz or greater though some cable operators are dragging their feet.

The process is quite elegant in practice. A digital cable ready product can access scrambled cable channels, as well as unscrambled digital and analog channels (but not pay-per-view or video-on-demand). To receive premium and other scrambled premium channels, the user inserts a security device known as a **CableCARD** (or **POD**—point of deployment—card) into a slot that could be built into a set-top box, DVR, or DTV. Most CableCARDs are deployed in boxes. Unscrambled digital channels are received by a **QAM** (quadrature amplitude modulation) tuner, which does not require the card.

To allay Hollywood's copyright-related concerns, unidirectional digital cable products must support some kinds of anti-copying technologies through their digital interfaces. HDTVs described as digital cable ready must have **HDMI** or **DVI** jacks with **HDCP** copy-prevention technology. Also, high-def cable boxes are required to have **IEEE 1394** jacks with **DTCP**. (The acronyms are defined later—see "Connections" at the end of this chapter.) The HDMI and DVI-HDCP interfaces completely prevent networking and recording, while the 1394-DTCP interface provides the option to restrict copying under some circumstances.

The unidirectional digital cable agreement provides for home networking and recording of digital video with some limitations on some content. One option that may be exercised by the cable operator on pay-per-view content is **copy never**. That may sound draconian but actually it does come with a loophole that would permit hard-drive-based digital video recorders to store paused programming for a minimum of 90 minutes. Another option is **copy one generation**, which allows material to be copied to a recording device (but not recopied). Up to 64 simultaneous first-generation copies are allowed—just don't try copying a copy.

In a temporary victory for consumers, one controversial feature that digital cable ready products initially weren't required to include is **selectable output control**, which would enable cable operators to use cable-ready gear to completely shut off recordable 1394 outputs and analog outputs including component video. Also prohibited—but only for un-encrypted broadcast channels—is the ability to **down-res** (reduce the resolution of) HD signals delivered to the component inputs of millions of early HDTVs. The down-res of other channels is not addressed in the agreement. This is how the signatories to the agreement summarized the current status of down-res for non-broadcast channels in a letter to the Federal Communications Commission: "The lack of such a provision should not be construed as an indication that down-resolution should or should not be permitted, but rather that the Commission should resolve this issue."

Hollywood lobbied the FCC for a waiver that would let selectable output control be used for video-on-demand delivery of major movie titles. The FCC eventually gave into this demand in 2010, providing a limited waiver that would allow use of SOC for 90 days or until disc release. The decision may be reconsidered after two years.

The initial digital cable ready agreement is for unidirectional (one-way) service and has been approved by the FCC. A newer type of cable card, known as **Tru2Way**, provides for **bidirectional** products—those with a return path, allowing two-way communication between the cable system and the box or set. Bidirectional products support additional services such as video on demand, interactive program guides, video games, and electronic commerce. Tru2Way has not been officially sanctioned by the FCC but has been adopted by several DTV and DVR makers including LG, Panasonic, Samsung, Sony, and TiVo. The cable companies are more open to the Tru2Way standard because it allows them to market new services. They've been less than enthusiastic about the unidirectional standard—despite having signed an agreement supporting it—and some consumers are having trouble getting old-style CableCARDs.

Meanwhile, CableLabs has come up with some cool new ideas for the next generation digital cable readiness beyond Tru2Way. One already implemented is Motorola's **M-Card**, which supports multiple-stream recording with a single card. Still in the lab is a means of supporting next-generation **IP-based video**, which would help the cable companies compete with the phone companies. Finally, the card may eventually give way to a chip with **downloadable encryption**. The FCC is promoting the idea of a universal adapter called **AllVid** that would embrace cable, satellite, and internet TV but technical details remain sketchy.

TV makers have offered digital cable ready products though their enthusiasm and support have waned over the years. They feel burned by the cable industry's halfhearted compliance (and frequent non-compliance) with the original CableCARD agreement. But, as discussed in the opening of this section, the FCC encryption ban may give digital cable readiness a new lease on life.

Tuner features

Picture-in-picture (**PIP**) allows simultaneous viewing of two or more pictures on one screen. You can monitor one channel while watching a picture inset of another channel or a rented movie. Most sets offering PIP display the extra picture using a separate tuner or another source such as a disc or tape. **Dual-tuner PIP** packs both tuners into one TV. Widescreen sets may have **picture-outside-picture** (**POP**).

Federal law requires that all TVs 13 inches and up have a **closed-caption decoder**. It displays printed material onscreen taken from the broadcast signal or from any captioned videotape or DVD. You may find additional captioned channels for foreign language, statistical displays, secondary analysis during news events, emergency announcements, weather, stock market reports, or program listings.

The hearing-impaired are not the only people served by closed-captioning. It's also helpful to those studying languages or improving their reading skills. It might even be helpful to you—yes, you—when you are watching a muted TV while talking on the phone, or someone is sleeping. It enables you to follow games or news events in noisy places like airports and bars.

Closed captions are part of a growing number of services called **Extended Data Services** (**EDS**). These might include elapsed time and name of program as you switch channels. They can set TV or DVR clocks, delay or extend recording when program times change, or display channel call letters and network affiliations.

Another federally mandated feature is the **V-chip**, which provides **parental control** and **channel lockout** functions. Concerned parents may block channels showing rated programs deemed unsavory for children. Should those same parents decide to unblock a channel at the midnight hour, they can restore a previously blocked channel with a numeric passcode.

Channel labeling liberates you from your cable system's meaningless channel numbering scheme, which may not even correspond to broadcast channel numbers. You can rename or renumber any channel to something that makes more sense to you, such as the station's call letters or network affiliation.

Auto channel programming scans through all channels to separate active from inactive ones. When surfers later step through channels, the tuner skips those with no signal, as well as any that have been blocked. It's not a bad idea to make a DTV set rescan channels on a regular basis to account for modifications in the broadcast signal. Station engineers are always tweaking something.

Automatic picture settings

More sophisticated sets have preset **picture modes** designed to enhance movie viewing, videogame use, etc. The movie mode—always preferred by videophiles—strives for a "filmlike" look. That might translate into lower contrast and brightness for viewing in a darkened room, and possibly a lower color temperature.

Picture memory is a useful feature that allows the user to store preferred picture settings for each input source. That way, if your DVD player is a bit on the greenish side, the set can compensate with a redder tint setting every time you select the DVD input.

Stereo and surround via broadcast or cable

The tuner's role goes beyond channel selection and picture. It also delivers audio, potentially including surround or stereo. A DTV tuner delivers audio encoded in **Dolby Digital** (see "Surround Sound/ Understanding surround standards"). If you want to receive analog stereo via cable—a must if you also want Dolby Surround sound—look for the clunky designation **MTS**, which stands for **multi-channel television sound**.

Surround and stereo subtly enhance everything from talk shows to sporting events. When the stereo soundtrack carries Dolby Digital, shows like your favorite primetime fare, Letterman, and *Saturday Night Live* take on a real you-are-there feeling, as though you were right there in the audience, in the heart of the crowd. Musical segments can sound almost as good as a compact disc.

SAP stands for **second audio program** and is found in all digital tuners and in obsolete MTS-equipped analog tuners. While not all broadcast stations or cable systems pass on these signals, some TV stations are becoming creative in their use of SAP. Some go beyond its obvious use for bilingual soundtracks to provide extra commentary during sporting events, uncut audio portions of R-rated movies, various kinds of background information, and audio-only news or weather services.

Other TV sound features

When TV makers load their sets with stereo or even surround speakers, the result is more a triumph of marketing than technology. Adding built-in speakers and amps without enlarging the set takes ingenuity—and sometimes the help of a high-profile partner from the audio industry—but these low-quality internal sound systems are rarely worth the extra cost. They're not good enough for movies, yet too elaborate (and possibly too boomy) for watching talking-heads television.

Unless you're planning a simple watchin'-the-news system with just a TV and video recorder, add-on audio components will do a better job of delivering movies. If you hook up your system right, you'll still be able to scale back to the TV's internal speakers when playing, say, the Weather Channel late at night.

Volume correction is a very helpful feature. Commercials are typically broadcast near the top of the allowable limit, unlike the programs themselves, which vary between loud and soft. A circuit that automati-

41

cally reduces the volume of blaring ads may save your remote-operating hand a lot of work. Dolby Labs, THX, and Audyssey have recently revived this old (but good) idea and license various forms of it for a/v products under the names **Dolby Volume**, **THX Loudness Plus**, and **Audyssey Dynamic Volume/EQ**. TV makers should seriously consider adding one of these processing modes to their sets.

But that may be unnecessary if the Commercial Advertisement Loudness Mitigation Act (or CALM Act) becomes law. This act of Congress would require the Federal Communications Commission to implement ad-volume rules already suggested by the Advanced Television Systems Committee. As this book was going to press, the CALM Act had passed both the House and Senate and was going back to the House for reconciliation before going to the president's desk.

Network features

Most major TV and Blu-ray manufacturers now include network features in the upper echelons of their product lines. Operating with a broadband internet connection, these features usually pull music, pictures, or video from the internet or from a router-connected PC. Web browsers may be included. Some brands—notably LG, Samsung, Sony, and Vizio—have also added onboard apps, some licensed from Yahoo, which provide news, weather, video, finance, and Flickr photo access. In Blu-ray players and Xbox gaming consoles, network connectivity can also be used to stream programming from Netflix, Blockbuster, or other sources. Adoption of HDMI version 1.4 and Adobe Flash will accelerate the spread of network features in DTVs by allowing devices to share an internet connection and exchange data in a standardized (as opposed to proprietary) manner. Now that viewers have become addicted to You-Tube and other internet video sources, network features are shifting from optional to indispensable.

Bedroom sets

A **TV/video** combo—usually a 27-inch-or-less TV with a built-in DVD player, VCR, or DVR—will rarely become the center of a viable home theater system. This is something you'd buy for a kitchen or a child's room. In fact, the controls are often simplified for a child's use.

Shopping for a bedroom TV gives you a chance to turn the set into a glorified clock/radio with AM/FM reception and **sleep/wake-up timers**. Even if you don't want to awaken to television, it can be comforting to set a sleep timer for, say, an hour. The TV will turn off automatically as you drift off to sleep.

DTV connections

Most readers of this book would be better off using a receiver for their switching needs. Therefore the big issue in TV connectivity is not the number of inputs, but the kind of video inputs available, which will influence the video performance of your home theater system.

- HDMI/DVI input
- HDBaseT
- IEEE 1394, DTV Link, FireWire input
- component video input
- RGB, RGBHV input
- VGA input
- S-video input
- composite video input
- RF input
- CableCARD slot
- front-panel video input
- digital audio input (optical or coaxial)
- stereo analog audio inputs
- audio/video outputs
- RS-232
- 12-volt trigger
- network connection

Two forms of high-quality digital connection have arrived on the back panel: HDMI and IEEE 1394. HDMI is the dominant one. To allay Hollywood's concerns, and to spur the growth of HDTV programming and disc releases, both come joined at the hip with digital rights management. Therefore HDMI comes with a form of DRM called HDCP, and IEEE 1394 comes with DTCP.

HDMI (High-Definition Multimedia Interface) carries both HDTV and multi-channel audio—although as initially implemented, only stereo capability may be activated, as technical protocols were still evolving. One attractive thing about HDMI is that the plug and jack take up a lot less real estate on back panels. HDMI was developed by Silicon Image though the HDMI Founders Group includes many major TV makers, movie studios, satellite TV operators, and CableLabs, the cable industry's R&D arm.

HDMI is an evolving standard that comes in several versions. Search spec sheets to find out what you're buying. The original **HDMI 1.0**, released in 2002, supports video and stereo audio, not surround. **HDMI 1.1** adds Dolby TrueHD, Dolby Digital Plus, and DTS-HD High Resolution Audio (new cutting-edge surround standards that are significant parts of the Blu-ray disc format) and DVD-Audio (a high-resolution music format). **HDMI 1.2** adds Super Audio CD, the main rival to DVD-Audio among high-res music formats, and is more computer-friendly. **HDMI 1.2a** add CEC (Consumer Electronic Control) which allows components in an a/v system to respond to commands in a coordinated way. Several manufacturers support CEC, often disguising it with phony proprietary names to take credit for something they didn't invent. **HDMI 1.3** came out in 2006 and supports higher bandwidth, greater color depth, a new mini-connector for camcorders, and automatic lip-sync to keep soundtracks coordinated with faces speaking on-screen. It also supports DTS-HD Master Audio, another new lossless surround format in Blu-ray. **HDMI 1.3a** makes small modifications in CEC, color depth, and other areas. **HDMI 1.3b** makes no functional changes for the consumer.

HDMI 1.4 brings many improvements. It adds an **HDMI Ethernet Channel**. That helps connected devices share an internet connection, and they can exchange data at 100Mbps, enough for any IP-based application. An **Audio Return Channel** reduces cabling by letting an HDTV pass the audio stream back to a receiver for surround decoding. Also supported is **4K x 2K** resolution (3840 x 2160 or 4096 x 2160), though right now it's used mainly in public theaters. There's also an expanded color space, support for nascent 3D media, a **Micro HDMI connector** that's half the size of the **HDMI Mini connector**, and an **Automotive Connection System**. HDMI 1.4 specifies three kinds of HDMI cable. See the "Cables" chapter or HDMI.org for details. Finally, the most recent version is **HDMI 1.4a**, which mandates a fuller set of 3D formats for movies, TV broadcasts, and games.

The more functional versions of HDMI—1.3 and up—are now entrenched. If all you want is top-quality video and audio, here's what you need to know about HDMI right now—get version 1.3 or higher. If you want 3D, make that 1.4a.

HDMI also now supports **latching connectors** which prevent the plug from falling out of the jack. These may use either a magnet or a mechanical prong.

The forerunner of HDMI is **DVI-HDCP (Digital Video Interface with High-bandwidth Digital Content Protection**), a variation of DVI-D plus copyright protection. You may see DVI's larger (much larger!) plug or jack on some video gear but it is on its way out. HDCP, the content-security scheme for DVI, also serves the same function in HDMI. DVI and HDMI can be physically bridged by an adapter—but the two connected devices will work together only if they share the same signal and protection protocols. That can be an iffy proposition.

Unlike IEEE 1394, discussed below, both HDMI and DVI pass a signal that is too fat to be used for digital recording or networking, and can't travel over long lengths of cable unless aided by a signal convertor.

Monster Cable, MF

New video interfaces improve picture quality by passing signals digitally. Cables pictured clockwise from top left: IEEE 1394, also known as FireWire, can be used for recording and networking. HDMI, like 1394, can carry both video and audio, but unlike 1394, it cannot be used for recording or networking. DVI is the forerunner of HDMI, but carries only video, has a huge unwieldy connector, and is being phased out. Bottom left, from top: 1394, HDMI, and DVI jacks.

What makes it so fat? Some describe it as an uncompressed digital signal—even I have made this error in past editions—but actually it does not reverse the effect of MPEG-2 video compression. It merely has a lot of extraneous data added to make it deliberately unwieldy. That hasn't prevented manufacturers from supporting HDMI or DVI. They are eager to embrace any new interface that would help them deliver higher quality signals with the blessing of the entertainment industry.

HDMI brings a couple of problems. It can't travel far, and the cables can be expensive. The brand-new **HDBaseT** standard would address those problems by translating HDMI to a format that can be carried by standard (and dirt-cheap) Cat5e or Cat6 cables with RJ-45 connectors. HDBaseT supports ultra-res high-def video up to 4K by 2K for distances up to 328 feet. Another major advantage is that the standard supports power as well as video connections. HDBaseT developers include LG and Samsung, so it might someday appear in those TV brands. Other developers include Sony Pictures and Valens Semiconductor.

Another digital interface known as **IEEE 1394** is being used in a few video and audio products. IEEE 1394 was named for the standard-setting body (and decree) that originated it, though it also has such proprietary names such as Apple's **FireWire** and Sony's **iLink**. The Consumer Electronics Association and the National Cable Television Association have agreed on the term **DTV Link** as well as the related terms **DVD Link**, **Web Link**, and **D Link** (the last one is for camcorders).

In one form, 1394 supports networking and recording of video signals—making it far more consumer-friendly in those respects than HDMI or DVI. It's also used by a few manufacturers (notably Denon and Pioneer) to connect universal disc players to surround receivers. In both spheres, implementation of IEEE 1394 has been stunted by the entertainment industry, which is militantly opposed to the possibility that digital copying may erode revenues.

For that reason, an early-generation DTV set-top tuner (the Panasonic TU-DS50) is in great demand because it's one of the few that can feed the IEEE 1394 inputs of a high-def video recorder without the hindrance of anti-copy flags. Unfortunately, that model has gone out of production, and the unfettered IEEE 1394 interface is available only in PC tuner cards. Most products with 1394 interfaces support **DTCP** (**Digital Transmission Copyright Protection**). 1394-DTCP has at least the nominal support of most major TV makers, the Consumer Electronics Association, plus the Warner and Sony studios.

Component video is the highest-quality analog video connection and the only one that's HD-capable. It connects a set to digital signal sources and usually takes the form of three red, green, and blue RCA-type jacks. It delivers video signals in three components (hence the name) consisting of a brightness signal and two signals representing color (or, more specifically, color difference—one color minus another). Be warned that not all component video jacks support HDTV. Some support only 480i signals. There are sets with one of each type—so check out the true capabilities of component video jacks if you want to hook up two different HD-component sources.

High-end projectors or multimedia monitors may include **RGB** jacks for connection to video processors or PCs. RGB jacks separate the signal into red, green, and blue and may use either screw-on BNC-type connectors or the more common RCA connectors. **RGB+HV** jacks add two more connections for horizontal and vertical sync. **VGA** allows connection of PCs.

S-video input, the second highest-quality analog video connection, uses a delicate multi-pin plug that separates the brightness and color portions of the video signal. Unlike component, it's not HD-capable. S-video is losing the competition for back-panel space in a/v products.

The third highest-quality analog connection is **composite video**, which mixes brightness and color together (hence the name composite). Like component video—which sounds confusingly similar—composite video uses a standard quarter-inch-wide RCA-type plug with yellow color coding and a fat pin at the center. It is not HD-capable. You're most likely to use it to connect VCRs and other legacy signal sources.

All TVs have at least one **RF**, or antenna, input. Instantly recognizable by its large threaded jack, this connection carries multiple channels of video and audio. RF is used most often for antenna, cable, or satellite input—if you have more than one of those, you may want more than one RF input—and is sometimes labeled "75 Ohms" (really old analog TVs may have a 300 Ohm antenna input instead). You'll also see that big RF screw terminal on DVRs, VCRs, cable boxes, and satellite receivers because, like your TV, each of these products includes a tuner (channel selector). If your DTV has no ATSC tuner, then you'll need a separate tuner box. If your DTV has a **CableCARD** or **Tru2Way** card slot, ask your local cable operator for the card, and then you won't need a cable box. Otherwise, connect your cable or satellite box to the DTV using the highest-quality output, preferably HDMI or component video.

Whenever you see **video** or **audio/video** inputs listed in literature, you can usually assume they're composite video, sometimes with S-video

47

added. **Front panel jacks**—convenient for connecting a camcorder or videogame machine—are usually composite or S-video. A set with surround decoding will have a **digital audio input**, either optical or coaxial. A set with surround or stereo capability also will have left/right-channel **analog audio/video inputs** to accommodate a stereo feed from a DVD player or stereo VCR (which often carries analog Dolby Surround). **Audio/video outputs** of various kinds are rarer but not unheard of. They provide extra flexibility by allowing you to feed picture and sound from the TV to other components.

Projectors and other high-end equipment are also likely to have a multi-pin **RS-232** jack, to interact with a touchscreen interface, and a **12-volt trigger**, to send control signals to other components such as projection screens. A **network connection** allows access to broadband-related network features.

The thorny subject of digital rights management

Any discussion of how sources connect to a DTV would be incomplete without mention of what you're *allowed* connect—and why.

The term **copy protection** begs the question of who is being protected—certainly not the consumer. There are a number of more straightforward synonyms including **copy prevention, copyright protection**, or just plain old **anti-copying**. The latest buzzphrase is **digital rights management**.

Congress first regulated digital media with the No Electronic Theft act of 1992, which mandated the use of the **Serial Copy Management System** (**SCMS**) in component CD-R decks (but not computer drives). SCMS (pronounced "scums") allows one generation of digital copying—the copy itself may not be copied. The No Electronic Theft Act of 1997 criminalized file sharing (including the nonprofit kind). The Digital Millennium Copyright Act of 1998 made *it a crime to tamper with copy-prevention schemes or to sell devices for that purpose. There are exceptions for libraries and researchers but the DMCA remains controversial among civil libertarians. Coming up are more draconian laws that would mandate the use of copy-prevention technology. In addition, the Supreme Court, in the now infamous Grokster Decision, has ruled that technology developers may be penalized for intent to cause copyright infringement, paving the way for a legal free for all.

A key battleground in digital rights management is the **broadcast flag**. The Federal Communications Commission approved it in 2004 but a federal court ruled against it in 2005. The broadcast flag's stated use is to prevent video from being shared on the internet, though it may limit consumer recording and networking options in other ways. There are 13 different licensees, each with its own application, including JVC (for D-VHS), Microsoft (for the Windows Media Player), Philips/HP (for their recordable DVD), Sony (for MiniDisc and Memory Stick), Thomson (for SmartRight home networking), and TiVo (for TiVo DVRs).

These different formats will impose various levels of restriction and how they will work in practice remains to be seen. Following, for the sake of background, is a theoretical overview of digital rights management from least restrictive to most restrictive.

- Copy freely
- Copy once
- Copy never, use only
- Use only with registration key
- Use only in approved regions
- Use with down-resolution
- Use never

Copy freely. Those who oppose any kind of copy prevention believe this is the natural order of the universe. Some products still work this way including certain CD-R drives, MP3 players, and cassette decks.

Copy once. The leading example is SCMS. Found in CD-R decks (but not PC drives), SCMS allows you to make a digital clone. However, you cannot digitally copy the copy. The limitation is built into the blank disc as well as the player—that's why black-box CD-R decks require special "music" discs that cost slightly more than ordinary "data" discs.

Copy never, use only. Successful examples include the Advanced Access Content System, used in Blu-ray; the Content Scrambling System (CSS), which ensures that only licensed DVD players can play DVDs; and the Macrovision APS (Analog Protection System), widely applied in VHS and DVD movie releases. VCR makers have cooperated with Macrovision by altering their products to work more effectively with APS. The process works by manipulating video signal levels and is designed to prevent only copying, not normal viewing, though it plagues certain video displays with picture bending or cycles of darkening and brightening. In a masterpiece of obfuscation, APS-treated software is sometimes labeled "Macrovision Quality Assurance"!

Use only with registration key. For example, a Windows PC logs onto Microsoft's server to authenticate your use of Windows. This happens once, then you're home free. Registration keys also figure in some CD copy-prevention schemes such as key2audio and MediaCloQ. These are rarely used now—most CDs are in the clear.

Use only in approved regions. Regional coding is built into most DVD-Video players and programs. It prevents releases in foreign markets from being illegally imported into other markets.

Use with down-resolution (or **down-res**). This alarming restriction would be on quality, not content. It would degrade the resolution of high-definition programming, rendering it standard-definition, through the component video outputs. Blu-ray carries the seeds of down-res through a digital flag called the **image constraint token**. However, the major studios have agreed not to implement the ICT in early titles, and it will not kick in until December 31, 2013, when it will become a requirement in all Blu-ray players made after that date. Down-res makes videophiles and audiophiles see red.

Use never. The leading example is the signal scrambling used by cable systems for premium channels and pay-per-view events. Of course you have no right to enjoy what you haven't paid for. Unfortunately, use-never status is also an unintentional side effect of some CD copy-prevention schemes that prevent CD-ROM drives from accessing copyrighted material. CD-ROM drives are used not only in PCs but in most CD players.

To be continued

DRM has become a red-hot issue among high-end videophiles. Those who voice their opinions in online message forums deeply resent the possibility that the Supreme Court's Betamax Decision, which legitimized video copying for personal use in 1984, may be rolled back in the age of digital television by corporate-technological fiat. Even more infuriating is the possibility that DTV sets now operating in homes might be rendered obsolete (at least without the addition of new set-top boxes and additional video processing that might degrade the signal). In the worst-case scenario, technology that prevents copying might even be used to prevent the *display* of HDTV.

However, without DRM, Hollywood is far less likely to release its crown jewels for HDTV consumption. Hollywood needs the revenue generated by its more successful movies for the same reason that I need the royalties generated by this book. Without compensation, creative work becomes impractical or even impossible.

The result is an impasse for DTV makers. If they don't implement some form of DRM, their DTVs won't be able to accept a copyright-protected digital cable or other signal. If they do implement it, consumers will have fewer DTV recording options (with HDMI and DVI—none).

As always, this story is *definitely* to be continued.

Shopping for a DTV

Digital television (or DTV) sets come in any of several forms including front or rear projection, direct view, or flat panel. Having decided that, you still need to think about where your programming will come from. You should do a lot of watching, and while you're doing it, you should be aware of limits—the limits of different kinds of displays, and the limits of what's being fed into them as you inspect them. Here, then, are some pointers for the shopper/survivalist.

Screen size and viewing distance

How big a screen do you need for home theater? Though viewers may prefer varying degrees of intensity, choosing the right screen size is not just a question of taste. The size and layout of the room should figure in your decision.

The traditional formula for relating screen size to viewing distance uses screen heights. Granted, this does not make for easy calculation, since manufacturers and retailers quote screen diagonal, not screen height. Anyway, measure the viewing distance from your favorite seat to the front of the screen. Screen height should be one-third of the viewing distance for digital TV (high-definition or line-doubled) and about one-fifth for analog TV. Or viewing distance should be three times screen height for digital sets or five times screen height for analog sets.

TV makers and retailers use diagonal measurements to fluff up the specs. A widescreen 34-inch TV actually has a screen height of 16.7 inches and a width of 29.6 inches. A nonwidescreen 27-inch TV is 16.2 inches high, 21.6 wide. Note that height in these two examples is nearly

51

identical—a 34-inch 16:9 TV is not much taller than a 27-inch 4:3 TV.

To calculate true picture height and width from the diagonal measurement, use this formula: In a 16:9 screen, height is 0.49 times the diagonal, and width is 0.87 times the diagonal. In a 4:3 screen, height is 0.6 times the diagonal, and width is 0.8 times the diagonal.

So if you're buying a 16:9 HDTV—the best choice for a home theater system—minimum viewing distance should be at least 1.5 times the diagonal. Or the diagonal should be no more than two-thirds of the viewing distance.

For a 4:3 HDTV display, minimum viewing distance should be at least 1.8 times the diagonal. Or the diagonal should be no more than half the viewing distance. (Note: I recommend buying 16:9.)

For a 4:3 analog or standard-definition display, minimum viewing distance should be three times the diagonal. Or the diagonal should be no more than a third of the viewing distance.

These are *minimum* distances. Some viewers may find them fatiguing and may need to sit farther back. On the other hand, if you have poor eyesight—or you're willing to trade a clean picture for emotional impact—you may be tempted to go for something larger, but be careful. When buying a digital set, you don't want to see the dotted **pixels**, or picture elements, that make up the picture. The best size is the one whose flaws are subjectively minimal when viewed at your preferred distance.

So get out your tape measure, figure out the viewing distance, and take that number to the store. Don't leave home without it! Any set you buy should look good from exactly that distance.

If you want a prospective TV purchase to fit into a piece of furniture, such as a home entertainment center, ask your tape measure to tell you whether the set's total height, width, and depth will fit. (And don't forget to take into account ventilation and the special needs of a set with side-mounted speakers.)

Know what you want

If you're buying a used TV (which I don't recommend) be warned that a real DTV is one compatible with the ATSC standard. A set with a **digital comb filter** or **digital effects** is not necessarily a true DTV. Nor is it safe to assume that any aging rear-projection TV is automatically digital. A digital HDTV will play any U.S. high-definition, enhanced-definition, standard-definition, or analog format.

Another digital-versus-analog distinction lies in the **tuner**, which selects channels. Before you order a high-def set, you also should decide what you're going to feed it. If you'll be getting video delivered over the air, you'll want a DTV (ATSC) broadcast tuner, either built in or out-board. Older sets may have only useless analog (NTSC) tuners or no tuner at all. Tuner-equipped displays may have the tuners integrated into the set or placed in a separate box. For satellite service, you'll need a satellite receiver, and a special type is required for HDTV. The same applies to HD cable or telco services and their respective boxes.

Know what you're seeing

Don't assume the pictures you see in a store are identical to those you'll see at home. For one thing, manufacturers ship sets, and retailers display them, with radically heightened **contrast and brightness** settings. At home, you should set them lower, both to adjust for less intense home lighting and to extend your set's life. Lower settings will take longer to wear out an LCD's backlight, plasma panel, direct-view tube, projector lamp, or other components with limited lifespans.

As discussed in the chapter on "DTV by the numbers," the DTV standard embraces several **display formats**, including high-definition and lower-resolution formats. Any digital display will convert incoming signals to its own native resolution. Try to get a sense of how a set looks when displaying, say, 1080i HDTV, versus 480p DVD. Some digital broadcasts and cable channels are standard-definition. So how an HDTV set converts and displays non-HD signals is significant.

A really good showroom may have access to better **video sources** than you have at home, such as a Blu-ray player. This tells you what a prospective set looks like at its best. A mediocre showroom may be simply feeding every set on the floor with analog cable. This tells you nothing at all.

Go ahead, watch some standard-def analog video. In the language of DTV, analog TV signals are 480i. A digital set will use a line doubler or a more sophisticated video scaler to upconvert it to its native resolution, such as 720p or 1080p. Many scalers are not very good, so you may see varying amounts of **motion artifacts**—distortion of moving objects or objects covered in a camera move. Look at what happens to moving objects, especially those with diagonal edges. The variation between sets, given the same material, can be telling.

Try some familiar Blu-ray discs or DVDs, preferably with a player that has a progressive-scan output. Some sets will display DVD in 480p. Others will upconvert it to 540p, 1080i, or 1080p, so you'll see more processing, but fewer scan lines. This kind of **scaling** is about to become part of your life, so learn what it looks like. The original DVD format is not high-definition television (HDTV), but it is a form of digital television with standard-definition resolution (SDTV). You may be watching a lot of purchased or rented DVDs with your new DTV.

If you're watching an HDTV source, find out whether it's 1080p, 1080i, or 720p. Then find out whether, and how, the set is converting it.

Watch a little of everything and get as much as visual experience as you can.

Limits of display tech

Sadly, not all sets described as HDTV can display all the resolution in a high-def test signal. Happily, this is because HDTV sets a high enough standard to challenge the very limits of display technology. So there are real differences in performance and you should be aware of them when shopping for a set.

Some (though not all) flat panels and microdisplay-based DLP and LCD projectors support HDTV. Others do not display HDTV except in an adulterated **downconverted** form. Any digital TV will convert all incoming signals to its native resolution. The shortcomings of LCD panels and microdisplays generally include poor **black level**—they reproduce black as dark grey. Some plasmas suffer from the **screen-door effect**—i.e. the pixels are large enough to be visible. Some LCDs suffer from **image lag**—things smear when they move. Any set with poor video processing will exhibit **motion artifacts** (see "Types of video noise" a few pages down).

Direct-view TV tubes cannot take full advantage of HDTV (though they often look dazzling even while taking partial advantage of it). All are limited in resolution by their **dot pitch**, the tiny holes in the tube's **shadow mask** that allow light to shine through colored **phosphors**. There's a tradeoff between resolution and brightness: the dots just can't be made small enough to produce high-definition pictures while also being big enough to allow adequate brightness.

Limits of picture and sound sources

One of the biggest mistakes you could make would be to buy a big

HDTV while ignoring what goes into it.

Now that videodiscs have gone high-def, the best signal source available to you may be the **Blu-ray** format. Most new titles now support 1080p, the state of the art in HDTV signals.

Over-the-air **HD broadcasts** are growing steadily. A strong signal can provide superb HDTV. Stations offering DTV channels deliver varying amounts of HDTV; the rest is SDTV. Analog TV, which survives in some cable systems, does not look nearly as good. To get broadcast TV, you'll need an antenna (either outdoor or indoor) and a tuner-equipped set. To get DTV, you'll specifically need a DTV tuner.

Satellite systems are by definition digital. However, many satellite channels are standard-definition, not high-definition. It may even be rather poor SDTV—heavily compressed, full of motion artifacts—due to the compulsion to jam as many channels as possible into the limited bandwidth afforded by satellites in orbit. Launching new satellites is, needless to say, expensive. You'll also need a slightly larger dish and an HDTV-compatible satellite receiver. Some of this may slow you down but don't let it stop you.

If you're dependent on **cable** service, ask a customer service rep for HDTV service (and an HD-capable convertor/descrambler). Also, be warned that there may be some confusion between HDTV service (using the 1080i or 720p formats) and DTV service (which may merely be digitized lower-resolution television).

The latest arrivals on the video-delivery scene are the telephone companies—**telcos** for short. **Verizon** is starting to offer **FiOS TV** in some (lucky) metro areas. FiOS stands for **fiber optics**, the technology Verizon is relying on to deliver the highest possible bandwidth, and therefore the highest potential picture quality. Areas not served by Verizon might get telco-TV service from **AT&T**. Its **U-Verse** uses a more economical combination of fiber and last-mile copper. Both companies are offering **triple play packages** combining television, internet, and phone service—in direct competition with cable operators.

Types of video noise

Part of being a videophile (or at least an informed consumer) involves being able to know when a display is doing a poor job and when it's merely the victim of a poor signal. That's why it's important to recognize different kinds of video noise, motion artifacts, and other distortions.

- block noise, macro blocking
- chroma noise
- contouring, posterization
- cross-color distortion, moiré
- dot crawl
- image lag, motion smear
- image retention, burn-in
- jaggies, stairstepping
- mosquito noise
- phosphor lag
- snow
- uniformity problems
- white crush

Block noise, or **macro blocking**, causes images to dissolve into squares of varying sizes—the worse the noise, the larger and fewer the blocks. It results from compression errors in the video signal and is usually not a fault of the display.

Chroma noise adds grain to broad areas of color. Analog source signals, or those delivered through composite video connections, have this problem.

Contouring, or **posterization**, roughens transitions from light to dark. It is often the fault of the display, especially plasmas.

Cross-color interference, or **moiré**, is a rainbow effect that results from color information being mixed with brightness information. It's usually seen in composite video connections. It should not appear in S-video, component video, HDMI, or other video interfaces when they are properly implemented—unfortunately, that's not always a safe bet.

Dot crawl causes zipper-like or checkerboard distortion along edges, especially vertical edges, and colored objects. It's a particular problem with analog video formats and composite video connections.

Image lag or **motion smear** is a problem with liquid-crystal-based panels and projectors. The crystals move when stimulated by current, but subside slowly. To minimize this problem, look for sets with a minimum 120Hz refresh rate and smoothing (or anti-judder) circuitry.

Image retention or **burn-in** occurs when graphics and other static images leave a visible impression on plasma panels and tube-based displays. Mild cases, at least in plasmas, may be reversed by running full-motion video for 24 hours. Severe cases are a permanent problem.

Jaggies cause diagonal lines to break into steps. That's why **stairstepping** is another name for this effect. The problem may be inherent in the signal or may be a fault of the display. Better displays use more sophisticated video processing circuits to eliminate this effect. The best way to gauge stairstepping is with an American flag fluttering in a strong breeze.

Mosquito noise blurs the outlines of sharp objects with shimmering noise. Like block noise, it's usually an artifact of video compression (not the display).

Phosphor lag causes the outline of a bright object to persist on-screen after it's supposed to have disappeared. It happens with plasmas and direct-view sets.

Snow is also called video noise though brightness noise or **luminance noise** would be more accurate. It is a grainy impurity embedded in the brightness portion of the video signal. A bad display may look noisy, but far more often, this is a fault in the signal.

Uniformity problems occur in LED-backlit LCD displays because of the way the backlights are deployed. Full-screen white or bright color is uneven across the surface, possibly in corners.

White crush occurs on displays with inadequate high-frequency bandwidth. It's usually not a signal-related problem.

Comb filtering

During the analog era, one of the most important video-processing circuits in better TVs was the **comb filter**, and even in DTVs, it has an impact on analog signals. Its function is to reduce video distortions that occur in NTSC, the analog TV standard, due to a kind of shotgun wedding. For greater efficiency in use of the over-the-air spectrum, the color (**chrominance**) part of the signal is interleaved with, and therefore interferes with, the portion governing brightness (**luminance**). Among the most objectionable flaws is **moiré**, often visible on a herringbone tweed or a football referee's jersey. (Check out the mess on Johnny Carson's tweed jackets in reruns of the old *Tonight Show*.) Another one is **dot crawl,** that annoying effect that mars the edges of objects (especially on sports or weather graphics). The best comb filter is none at all. That's why component video and S-video are superior forms of analog video connection—there's no need to separate color from brightness because they're not mixed together in the first place.

A digital video scorecard

In the digital television era, there is a pecking order in picture quality. Home theater buffs with the biggest screens have the biggest motivation to pay more for a better picture. After all, a screen big enough to capture your imagination is also merciless enough to magnify every flaw in the picture source that feeds it.

Following is a scorecard of video formats in order of resolution. Since half of these formats are digital and half are analog, the methods of specifying resolution vary. Digital formats usually specify resolution in a simple grid of pixels, vertical by horizontal. For analog formats, the first number is still vertical resolution, but the second is lines of horizontal resolution, or TV lines (TVL). For a refresher on this knotty subject, reread "DTV by the numbers/Interpreting TV specs."

1080p HDTV (1080 x 1920 pixels, progressive): This Holy Grail of DTV formats has more than a thousand lines of vertical resolution. That's the number of horizontal scan lines that can be counted vertically from screen-top to screen-bottom, though the whole point of HDTV is that they're so small, they're hard to see. This progressive scanning format (that's the "p" in 1080p) scans the whole picture in one pass, like your computer monitor. Unfortunately, the 1080p signal is too big to travel through a standard-sized 6MHz broadcast channel using currently available MPEG-2 video compression. Until recently 1080p was mainly an archival format. However, a rapidly increasing number of HDTV displays now convert all incoming signals to a native resolution of 1080p as a picture-enhancement strategy. True 1080p signals are available via Blu-ray disc and are starting to become available via satellite, cable, and telco. However, there are no 1080p over-the-air broadcasts.

1080i HDTV (1080 x 1920 pixels, interlaced): There are plenty of sets with more than a thousand scan lines, backed up with an increasing array of broadcast programming. Many of these sets use a controversial technology called interlacing (that's the *i* in 1080i) to scan the picture in two interlocking passes. High-end videophiles say this is a bad thing, leading to distortions of moving objects called motion artifacts. Others retort that the effects are subtle and 1080i looks beautiful (and compared to analog video, it does). This form of HDTV is available from broadcast networks including CBS and NBC and carried by cable operators and telcos. It's also offered by the Dish Network and, in slightly adulterated form, DirecTV. Whether you opt for broadcast, satellite, cable, or telco HDTV delivery, you'll need both a set-top box to decode the signal and an HDTV set to display it.

768p HDTV (768 x 1366 pixels, progressive): Used in some LCD HDTVs, though there is no 768p signal format per se in television. All incoming signals are converted to the set's native resolution.

720p HDTV (720 x 1280 pixels, progressive): Favored by ABC and Fox—and also available via broadcast, satellite, cable, and telco—this form of HDTV has fewer scan lines (you guessed, 720) but does use PC-friendly progressive scanning. The 720p format figures prominently in more affordable flat panel displays. In many DTVs, the 720p signal is upconverted or downconverted depending on the set's native resolution.

480p EDTV/DVD (480 x 720 pixels, progressive, 540 TVL): Even at enhanced-definition resolution, digital video can look great. The best way to see it is on DVD—preferably feeding a progressively scanned signal from a high-end DVD player to a DTV. Thus avoiding interlacing, the videophile gets eye-popping color (especially red) and a very clean picture, even though it's not as sharp as HDTV. Though the 1080i and 720p HDTV formats are increasingly dominant in broadcasting, some digital broadcasts are in the 480p EDTV format. While DVD has 720 horizontal pixels, if you use the old analog-TV yardstick, lines of horizontal resolution (TVL) are 540 or less.

480i SDTV/DVD (480 x 720 pixels, interlaced, 540 TVL or less): Low-end DVD players deliver an interlaced picture. The interlaced (480i) format is a step down from the cleaner picture possible with progressive DVD (480p). However, it still beats analog broadcast, cable, and VHS sources hands down, and probably will outperform most satellite signals. (See "Picture & Sound Sources/Disc players" for more on the difference between 480p progressive and 480i interlaced DVD.)

480i SDTV/satellite (480 x 704 pixels, interlaced, 480 TVL or less): Most satellite broadcasting has the same number of lines as DVD. However, most satellite receivers use interlacing, diminishing picture quality to about the level of good analog television. By stuffing their systems with as many channels as they can, satellite operators further reduce picture quality, leading to blocky pixellated distortion. This has done little to diminish the popularity of satellite service though it also probably has fed much of the enthusiasm for DVD. Lines of horizontal resolution (TVL) measure at 480 or less.

An analog video scorecard

At this point we move from digital video, measured in vertical by horizontal pixels, to analog video, where horizontal resolution is measured only in TV lines (TVL). All of these formats are interlaced.

Laserdisc (480 x 425 TVL): The 12-inch precursor to DVD delivered nearly the same resolution as its five-inch replacement. Unfortunately, Pioneer stopped making LD players in 2009, though you can still buy them used. If you look hard, you might find a top-line surround receiver or preamp/processor with an RF-demodulated output to accept Dolby Digital from laserdiscs, though digital surround is generally a privilege of the digital-age videophile.

Super VHS (480 x 400 TVL): While S-VHS delivers a picture with slightly more than 400 TV lines, approaching the resolution of laserdisc, its picture is full of noise. That's because it excels over VHS only in the recording of the brightness (luminance) portion of the video signal—the color (chrominance) portion is no better than regular VHS.

Cable (480 x 300 TVL): Cable operators can and do deliver HDTV and SDTV. But some channels in some systems are still analog. Cable ops get some signals via satellite, and others by special phone line, but it's not unusual for them to grab some channels the same way you might—using an antenna. Either way, these signals then have to travel through miles of wiring, hitting a few video amplifiers along the way, until they get to your home. A newly and carefully rebuilt system can still provide decent quality—my cable signal improved noticeably when Time Warner pulled the decades-old cable out of my building's stairwells and replaced it with a new, better-insulated system properly installed in conduit. And then jacked up the rates, of course.

VHS (480 x 250 TVL, interlaced) **and D-VHS** (up to 1080 x 1920 pixels, progressive): VHS is both the worst and the best of all possible worlds. It is limited to 250 TV lines with a softening of the picture that's easily visible even on a 20-inch set. But if you're one of the lucky few with a digital D-VHS recorder or HD-enabled digital video recorder, you can record HDTV. This terrifies Hollywood, so you know it must be a good thing.

Surround Sound

*A*fter the big picture, big sound is the other major requirement of a home theater system. Like a big screen, surround sound is designed to engulf the senses, suspend disbelief, and pull the audience into the story. In a good home theater, the quality of surround sound can easily surpass anything available in your local cineplex. You have more control over the level of sound, which is abusively loud in today's moviehouses, and the quieter moments of a movie will have greater impact without the senseless yapping of today's horrendously rude cineplex audiences.

Surround speakers

Speakers are among the most important components in any system, whether for movies or music, in surround or stereo. And sometimes the bulkiest. Human hearing being variable, they're perceived in different ways by different listeners. That's not just a matter of specs and acoustics—musical tastes and hearing capacity also play a role (the young hear better). However, once you've found a set you like, they'll serve you for at least a decade (or as long as the drivers and the parts surrounding them hold out). So invest in the right speakers.

A typical surround sound installation includes a **5.1-channel** (or more) array of speakers driven by a receiver. As explained way back in the introductory chapter, that usually includes three speakers in front (left/center/right) and two toward the rear of the side walls (left/right) plus a subwoofer (or two). Some systems have three front speakers, two side-surrounds, and one or two back-surrounds in a **6.1-** or **7.1-channel** configuration though in my opinion few rooms need that much surround coverage. Recently, height or width channels have substituted for (or supplement) back-surrounds.

Within that basic multi-channel array are many possibilities. You might anchor the system with large front left and right speakers and dispense with the sub. Those left and right speakers may have built-in subs of their own. You might have five identical satellites and a subwoofer (as my system does), or four identical sats, a horizontal center design, and a sub. You might prefer your center speaker to be a horizontal design or a duplicate of the front left and right. Purists prefer the center speaker to be a perfect duplicate of the other front speakers. If a center speaker isn't convenient, you might eliminate it entirely and run the surround processor in phantom center mode. You might prefer bipolar or dipolar speakers in the rear. You might have one, two, or no back-surrounds, or height speakers, or width speakers. There are many variations.

Regardless of what you choose, it's important to match the front and rear speakers to produce a solid **soundfield**. Even if their enclosures are not the same size, try to match their driver sizes, so that they'll all deliver **directional** information (front/back, left/right) in a uniform way. That's called **timbre matching** and it generally points toward a one-brand speaker system. However, don't assume all speakers from the same brand have the same tweeter or crossover or voicing. Check the specs—and listen.

The larger the room, the bigger your speakers should be. Their job is usually described as "moving air"—a physical action—though it would be more precise to say that speakers create changing patterns in air pressure like ripples in water. Bigger drivers in larger enclosures can do more of this. A bigger room also requires more bass energy, so bigger speakers might be worth the investment in space and money.

Main left/right speakers

Within every surround system is a vestige of the old stereo pair. Front left/right speakers can take many forms. They may be the same size or larger than center and surround (rear) speakers, especially for listeners who prefer using large speakers to play in two-channel mode (**stereo**). For a uniform surround soundfield, it's generally best for the center speaker to resemble the others to the extent possible, sharing tweeter

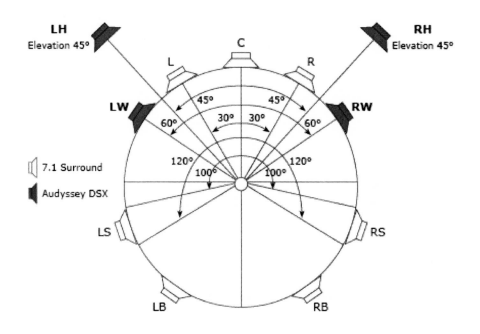

Audyssey

Surround sound's standard speaker configuration is a 5.1-channel array consisting of front left, center, and right speakers, plus side-surround left and right, and at least one subwoofer. The sub is not shown in this diagram. Note that the three front speakers (L, C, R) form an arc along the front wall, with the side-surrounds (LS, RS) toward the rear of the side walls. This is all you need for surround sound, movies or music. However, some 7.1-channel systems add two back-surrounds (LB, RB) along the back wall. Others add front height speakers (LH, RH) or front width speakers (LW, RW). Using all these options plus a couple of subs can result in a system with up to 11.2 channels, but for home use, the 5.1 standard still provides excellent room coverage, and you shouldn't feel pressured to go beyond that. If you really want to go beyond 5.1, start by adding an extra sub. This diagram from suggests optimum speaker placement with the front left and right 30 degrees off center, front height 45 degrees out and up, front width 60 degrees out, and side-surround 100-120 degrees out. (Ccourtesy of Audyssey, following the ITU-R BS.775-1 standard.)

sizes and materials. Drivers that handle midrange/woofer duty should be the same or very similar in construction.

Floorstanding speakers are usually big enough to produce serious bass. You don't need big speakers to put together a good home theater system, especially if the system includes a subwoofer; many speaker makers even recommend that you cut off low bass going to the satellite speakers and route it entirely through the sub. But many audiophiles still prefer the old method of using two bigger speakers, both for music reproduction, and for better bass coverage in general. You might consider buying big main speakers *instead* of a subwoofer (surround 5.0), setting the surround processor to route all bass through the front left/right speakers. Going a step further, you might even eliminate the center speaker by switching your surround processor to the phantom center mode, which routes the center-channel signal to the left and right speakers. Look for slimline designs with big woofers limited to the side of the enclosures. Add internal amplification to power those side woofers and you've moved into the next category.

Powered towers are extremely practical. First, they make it easy to eliminate the separate subwoofer from the system—your spouse will like that. As a bonus, they add *two* built-in powered subwoofers to your system. *You'll* like that! More benefits: The designer will make sure that the subwoofers blend well with midrange and high-frequency drivers. And with the speakers' internal amplifiers driving the subwoofers, your receiver may be less likely to distort, giving the whole system more dramatic dynamics and a sense of ease. Only a few speaker models use this configuration, but for bass addicts, it may be the best way to go. Keep in mind that you'll need to run not only a speaker cable to each speaker, but a power cord as well, to feed the internal amp.

Bookshelf speakers (or **near-field monitors**, to use the recording studio term) are also an excellent foundation for a home theater system. The name is somewhat deceptive—most speakers aren't designed to sound their best on shelves, where vents can be blocked, and acoustic conditions are poor. But many *bookshelf-sized* speakers have just enough bass to function without a subwoofer, so you can save your pennies and buy a really good subwoofer later to give the bottom octave that final degree of authority. They also happen to sound excellent in stereo, often outperforming more costly floorstanding models in overall accuracy.

LCR speakers typically have a woofer-tweeter-woofer array and can be used for all channels, or just the front three channels, or just as a horizontal center. Note that the dual-woofer array in a horizontal speaker can cause sum-and-cancellation effects called **lobing**, which is audible as uneven midrange and bass response depending on where you sit. To avoid this, some manufacturers add a midrange driver, which does a better job of anchoring voices. A few others will vary the crossover circuitry between the woofers and tweeter.

Satellite & subwoofer sets replace big speakers with small satellite speakers that get a boost in the lower octaves from a subwoofer, which may or may not be **active** (self powered). They're a popular way to introduce surround sound into a room that hasn't got space for boxy conventional speakers. Depending on their size and bass response, the sats may or may not function independently of the sub. That's fine since most surround receivers let *you* decide how bass is divided between sats and sub. The smallest sat/sub sets function best in a small to medium-sized room. Because they lean heavily on the sub—which must produce a wider range of frequencies to compensate for the small size of the sats—there may be some unevenness where the bass meets the midrange, indicating poor integration of sats and sub. However, the best mini-sat/sub sets can and do sound good.

The center speaker

The front center speaker—the anchor of a 5.1-channel or more surround system—is not without its controversies. Everyone agrees that the center channel is where dialogue resides and that it plays a significant role in delivering special effects and—with concert videos, high-res music formats, or stereo sources adapted to surround—even music. After that, opinions diverge.

Some high-end surround buffs insist that nothing less than three identical front speakers will provide the best results, pointing to acoustic problems associated with center-speaker designs using a woofer-tweeter-woofer configuration. In some horizontal centers, as noted above under "LCR speakers," the two woofers may cancel each other out at certain frequencies. However, many speaker makers and consumers are equally comfortable with (or resigned to) horizontal front center speakers.

Surround and back-surround speakers

The term **surround speakers** refers, somewhat confusingly, to the rear speakers (as opposed to the whole set) because they provide the surround effects. In 6.1- and 7.1-channel systems, there are surround speakers against both the side and back walls.

The main choice in surround speakers is between **direct radiating** (read: conventional) speakers and the trendy **bipole/dipole** types that have drivers on two sides of the enclosure. The front and rear drivers in **bipolar** speakers operate **in phase** with one another—that means they move in the same direction. **Dipolar** speakers operate **out of phase**. Some speakers can be switched to operate in either mode. A few can also switch off one set of drivers to function as **monopole** (conventional) speakers. Especially with dipoles, the main effect is a more subtle and diffuse feel in the surround effects. The drivers fire to the front and rear of the room, not at the listening position, to keep surround effects from becoming an unwelcome distraction.

In 5.1-channel systems, a single set of surrounds operates on the back of the side walls and they are referred to, simply, as the surrounds. In 5.1+ systems, they are called **side-surrounds**, while two additional speakers on the back wall are called **back-surrounds**. A 7.1-channel system uses two back-surrounds; a 6.1-channel system uses only one. The addition of back-surrounds is an idea that made sense in public movie theaters, where it originated as a way of spreading surround effects evenly through the rear of the auditorium, but then it was hastily ported to home theater without any consideration for practicality. The rush to 5.1-channel-plus formats is not a good move for anyone whose screening room is less than a cavern.

In 6.1- and 7.1-channel systems, THX recommends bipole/dipole speakers in the side-surround positions and direct-radiating speakers in the back-surround positions.

Height and width speakers

In 2009, **Dolby Pro Logic IIz** and **Audyssey DSX** introduced the concept of height channels, while DSX also introduced width channels. Height speakers are hung above and outside the main front left and right speakers. Width speakers go just outside the main left and right speakers. Because these new listening modes deliver fairly amorphous information, small satellite speakers will suffice for them. They are explained in more depth in the chapter on "Understanding surround standards."

The subwoofer

In either surround or stereo, a subwoofer (or two) can make a big

PSB, Boston Acoustics, Infinity, Paradigm

Clockwise from top left: The PSB Synchrony Series, including the Synchrony Two shown ($3000/pair), earns its high-end pricetag with high quality of construction including seven-laminate side panels and extruded-aluminum corners. The Boston Acoustics VS 240 monitor makes a high-performing foundation for a 5.1-channel satellite/subwoofer set with its multi-layered pressed-wood panels and unique bell shape. The center and sub are the VS 325C and VPS 210, totaling $3695 for the full set. The Infinity PS-212W subwoofer ($679) uses 2.4GHz wireless technology to make placement a cinch. Finally, the Paradigm Signature 2 v.3 is the third-generation stand-mount model in the Canadian brand's top-of-the-line range. It is unfailingly neutral and sounds good both on- and off-axis. One of its older cousins in the Paradigm Reference line is the author's reference speaker.

difference. **Driver size** may be 8, 10, or 12 inches (occasionally more).

A subwoofer may be a simple device that uses a single 10-inch driver, something economical in terms of both space and money. Or it might be a large heavy object with multiple **active drivers** and/or **passive radiators**, the first powered by internal amplification, and the second by the powerful movement of air within the enclosure.

Smaller subs are more suitable for most apartment dwellers. With rare exceptions, a driver size lower than eight inches disqualifies the device as a subwoofer, regardless of what the manufacturer calls it, and qualifies it as a woofer. Such un-subwoofers are seen mainly in low-end compact systems.

At the other extreme is the driver that's so big that the amp has trouble starting and stopping its movement. It doesn't help that the struggling amp may be deceptively specified at its peak (rather than continuous) power rating, which is legal, because amps built into subwoofers aren't covered by Federal Trade Commission regulations. Even so, a large room can benefit from the muscle of a well-designed 12- to 15-inch sub. Just don't get excited and buy a bad one.

If in doubt, buy a simple 10- to 12-inch model with at least one driver facing you and nothing more than a thin speaker grille in between. That setup will deliver low bass in roughly the same time domain as the other frequencies, keeping the sound in focus. Such **front firing** designs produce the most accurate bass, though that dry description hardly does justice to the indescribable feeling of big bass waves hitting your body. **Bottom firing** designs achieve much of the same effect.

Subwoofer designs are diverse enough that drivers may face any direction. In fact, they may sit entirely within the enclosure, emitting sound through a **vent** or **port**. There might even be tubing involved. These **bandpass subwoofers** are efficient and may come with impressive specifications. Unfortunately, listeners pay the price in muddier and less accurate bass.

Most subs come with both a **high-pass filter** and a **low-pass filter**. The high-pass filter is generally used with a speaker-level connection (i.e. speaker cables) while the low-pass filter is generally used with a line-level connection (lower-voltage interconnect cables). The high-pass filter is sometimes fixed at a certain crossover frequency, or may be affected by the crossover control. It enables the sub to strip out the lows and pass the rest of the signal to the satellites. The low-pass filter is variable, adjustable via the sub's crossover control, and limits the sub to sounds below the crossover frequency. A surround processor contains a low-pass filter of its own that duplicates the one in the sub.

For that reason, a desirable feature found in very few subs is a **low-pass filter bypass**. It eliminates the sub's low-pass filter and lets the receiver or surround processor determine what bass frequencies go to the sub. By eliminating the sub's filter when it's not needed, the bypass results in cleaner and more accurate bass. Sometimes an input labeled **LFE** (low frequency effects) accomplishes the same thing, bypassing the sub's filter.

Subs with servo circuits can track and control woofer movement with great precision. Infinity invented this kind of sub amp. (I have fond memories of an Infinity HPS sub that scared the devil out of people on the opposite side of the testing-room wall. They came in and asked, "Are you OK?") Paradigm, Velodyne, and a few other brands offer something similar.

A rare but highly desirable feature in subwoofers is **equalization**. An equalizer is a glorified tone control. In a sub it smooths out frequency-response humps that occur when bass waves interact with the room. EQ may be either **graphic** (fixed frequency bands) or **parametric** (variable frequency bands). The kind built into subs is usually parametric. Some subs with EQ (notably Infinity and Paradigm) come with a kit—containing a sound meter, test CD, and other paraphernalia—that allows you to measure the room and dial in the best settings.

In-wall and in-ceiling speakers

It's possible that you—or a loved one—may not want to live with a home theater system that resembles Easter Island. If your spouse resents having all those obelisks cluttering up the room, the use of in-wall or in-ceiling speakers may make the difference between having a surround system and not having one. The good news is that in-wall speakers are getting better and better—some perform nearly as well as the best conventional ones. Here are just a few things you'll need to know.

In-walls (let's just call them that) are as subject to the laws of acoustics as any other kind of speaker. It's even more important to get the placement right because you're making a hole in the wall or ceiling. For that reason I strongly recommend that you seek the services of a well-qualified custom installer to get it right. (See the chapter on "Connecting a Home Theater System/Hiring a custom installer.")

To provide a little wiggle room, there are some in-wall designs that use a pivoting tweeter, which can be moved around to change the dispersion pattern. However, woe be to the consumer whose in-walls just aren't in the right places.

In-ceiling speakers suffer the acoustic disadvantage of having to fire at the floor. In bare-floored rooms this becomes especially problematic. While you wouldn't want to use them for front speakers in a surround system, they might be a helpful problem solver in the rear positions. Look for in-ceiling models with angled, pivoting, or motorized drivers.

For best acoustic results—especially with an in-wall sub—you'll need to install a **back box** behind the speaker to control resonance. Otherwise the wall becomes a resonating chamber. Result: bad sound.

On-wall speakers

The popularity of flat-panel video displays has spurred the demand for flat-panel speakers that wall-mount with **keyholes** or **threaded inserts**. Some sound better than others but all operate under a handicap.

Simply placing conventional speaker drivers into a flat enclosure is not a recipe for good sound. The enclosure is a critical component of a speaker, and without enough physical depth, a speaker cannot produce much sonic depth.

Wall placement brings further acoustic problems. The closer the speaker is to the wall surface, the higher in frequency (and more objectionable) its interaction with the wall becomes.

Alternatives to on-walls include in-walls, which don't interact with the wall surface at all, because they're behind it, *in* the wall. Also consider small satellite speakers on slender stands—they can be both visually nonintrusive and great-sounding.

Bar speakers

This emerging speaker category, intended to complement flat-panel displays, comes in two basic configurations. One type is a three-channel speaker that handles the front left, center, and right channels, leaving the surround channels to other speakers. The other type usually tries to simulate surround sound from a single horizontal speaker. It may use either legitimate Dolby/DTS surround processing or, more often, various faux-surround modes. While there are some high-end implementations worthy of a decent bedroom system (from Polk, B&W, and Phase Technology) bar speakers in general are not suitable for a primary home theater system.

Exotic speaker types

While speaker technology is staid and stodgy, there are a few alternative technologies to liven things up. One is the **horned speaker**, which places the tweeter deep into the enclosure, in a recess designed to shape the flow of sound in a specific pattern. This maximizes loudness in the listening positions covered by the horn's dispersion pattern. That in turn makes horned speakers the most efficient type, allowing your surround receiver to achieve higher volumes during peak musical and cinematic moments. It also introduces some cupped-hands coloration, though an experienced designer can minimize this problem. The acknowledged masters in this area are Klipsch—whose founder, Paul Klipsch, is remembered as the father of the horned speaker—and JBL, which has been selling large pro-level horned speakers to movie theaters practically since the dawn of the talkies. In more conventional speakers, tweeters are often slightly recessed into **waveguides**, which are distant cousins of horns.

Electrostatic speakers are a legitimately high-end flat-panel speaker technology. Electrostats replace conventional cone- or dome-shaped speaker drivers with a diaphragm that operates in a charged field. This eliminates any need for sound-polluting crossover circuitry between drivers. And it provides swooningly realistic high frequencies—the increased clarity can be both refreshing and addicting. However, low frequencies don't fare as well, so most electrostats are best mated with a subwoofer to beef up the bass. Also, some electrostatic models require a direct power connection, so if your room doesn't have enough AC outlets, add your local electrician to the list of ancillary expenses.

Another form of flat-panel speaker technology hails from England where **NXT** has pioneered the use of flat membranes—made of various plastics and even glass—in lieu of conventional woofers and tweeters. The panels are stiff, but light, and excited by a simple motor-like device just as you can excite a tabletop by rapping your knuckles on it. Eliminating conventional cones and domes provides more even coverage. And again, there's no need for a crossover. There are some sonic disadvantages too: The panel may add coloration of its own. And bass is on the lightweight side. So far NXT has been used mainly in small or portable devices.

Flat diaphragms stimulated in the conventional fashion—by electromagnetic voice coils—are also appearing, notably in speakers from the Harman International stable (Infinity, JBL, Revel).

71

Interpreting speaker specs

- drivers
- voice coil
- enclosure: vented or sealed
- two-way, three-way
- driver types: woofer, midrange, tweeter
- driver materials
- driver sizes
- crossover
- power handling
- impedance
- sensitivity, efficiency
- frequency response
- dispersion
- shielding
- matching center, surround

The best way to buy a set of loudspeakers is to take over the store, throw out all the staff, acoustically treat the listening room to eliminate all echoes, bring in your own reference amps, and plot response curves from test signals. In real life, the best way to find good speakers is to listen to a lot of them before buying (or get a money-back guarantee from manufacturers who specialize in online sales). Before you go shopping, it doesn't hurt to learn a little about what speaker specifications mean. In fact, you can look up many of these specs on the websites of speaker makers.

A loudspeaker is a box containing one or more **drivers** (paper, plastic, metal, or composite **cones** or **domes** in metal **baskets**) that emit different frequencies of sound when stimulated by electrical signals. An electromagnet called the **voice coil**, suspended in a magnetic field created by other magnets, is connected to each driver and controls its movement. **Crossover** circuitry determines which frequencies, low or high, go to which drivers. The drivers "move air" in the room: they create the variations in air pressure that we perceive as sound. Now, how does that translate into specs?

Start with the **enclosure**. As a practical matter, is it too big for the room? The type variously known as **vented**, **ported**, or **bass reflex** contains a hole in the front or back to leak a little more bass into the room. Slightly rarer is the **sealed** or **acoustic suspension** type, which may be

72

more accurate if the designer has a good ear. All other things being equal, a vented design will play deeper, but watch out for muddy-sounding peaks in the midbass response, as well as audible port turbulence (**chuffing**). A well-made speaker will have a **braced enclosure**, with structures inside the enclosure that prevent it from resonating. The enclosure itself should vibrate as little as possible—unwanted resonance adds undesirable coloration to the music, especially in the upper bass frequencies. A good enclosure does not have a note of its own; when you rap on it with your knuckles, you should hear a dull thud, not a discernible pitch.

In a **two-way** design, the drivers typically consist of a **woofer**, which reproduces lower sounds, and a **tweeter**, which reproduces higher sounds. **Three-way** designs add a **midrange** driver and a second crossover. More drivers may or may not produce better sound—again, it's up to the designer. Rare **coaxial** speakers place the woofer and tweeter in the same axis (one inside the other). That produces better imaging, and more accurate sound over a wider area, because the listener is always the same distance from both drivers. A possible downside is cupped-hands coloration though it may be overcome by sophisticated design.

Stick to **woofer sizes** greater than 5 inches if you want to avoid total dependence on the subwoofer for bass response. Avoid woofers greater than 10 inches unless they're built into a subwoofer or mounted on the side of a tower.

Driver materials vary in the maximum speed at which they vibrate, something particularly critical in the tweeter. The fastest materials are metal (aluminum or titanium), though good-sounding drivers may also be made of plastic (usually polypropylene), textiles (including silk dome tweeters and bulletproof Kevlar woofers), paper (often treated or "doped" with something to enhance sound or durability), or various composites and laminates (like the aluminum/ceramic composite made and invented by Infinity). Faster tweeters are not always better. Misused, metal tweeters can be harsh and ringy. Paper, the cheapest material, can sound excellent. Also pay attention to the driver surrounds (which fasten the driver into the basket) and to the basket itself (diecast metal is better than pressed metal).

Debating the fine points of **crossover** design is beyond the scope of this book. But be advised that while there are excellent speakers with crossovers between the midrange/woofer and tweeter at anywhere from 1.8kHz to 4kHz, 2kHz is believed to be a significant number. Avoid tweeters crossed over below 2kHz (they're likely to distort) or 8- to 10-

inch woofers crossed over above 2kHz (they're likely to create a cupped-hands effect with voices).

Manufacturers may specify **power handling** as a **range**, a **peak** number, or an **average** number. Does that mean lower-powered amps are easier on your speakers? Not necessarily. A struggling amp may generate harsh distortion that hurts both your speakers and your ears. More power is better than less. Clean power is better than dirty power. This will come up again when you buy a receiver or power amp.

Your receiver or power amp, or its manual, contains clues to what kind of speakers you need. Speakers are rated for **impedance**, the amount of impediment a speaker puts up to an audio signal entering through its terminals. Think of lowering impedance as turning on a faucet: the more you open the faucet, the more water runs through. Likewise, as impedance drops, the speaker's demand for current rises, and the amplifier's workload goes up. In speakers, an average impedance is 8 Ohms, which is easy for any amplifier to drive. Impedance of 6 Ohms is a little harder, 4 Ohms harder still, and 2 Ohms or less requires high-end heroics. Speakers operate at a range of impedances that varies according to frequency—in other words, a loud, percussive bass drum thud might have a lower impedance than a whispery flute solo, and therefore would exert more strain on the receiver. Many inexpensive receivers will not comfortably drive anything much lower than 8 Ohms, so examine this spec carefully when mating any speakers with any amplifier.

Sensitivity is an absolutely crucial spec which may be given confusingly. Specified in **dB** (**decibels**), usually using a one-watt test tone measured one meter from the speaker, it tells you how loud the speaker will play in a non-echoing (**anechoic**) test chamber. An alternative version of this spec is **room efficiency**, which substitutes an average room for the non-echoing chamber. When you see that term, allow for rating inflation of 2-3dB. Otherwise consider 88dB about average. Lower sensitivity figures are forgivable in high-end speakers, assuming you have powerful amps to drive them. Higher ones suggest that the speakers will provide extra volume capability, though possibly at the cost of accuracy or listening comfort. Sensitivity directly affects how much power your system needs to perform well. Never forget that that every 3dB reduction in sensitivity requires a doubling of amplifier power to achieve the same volume.

Frequency response attempts to capture the speaker's distribution of low, middle, and high sounds in a few numbers that may be helpful or highly misleading. The compact disc format is designed with a frequency response of 20Hz-20kHz (20 Hertz to 20,000 Hertz, or vibrations per second). Human hearing starts lower but rarely goes as high. If a speaker measures at 40Hz-25kHz +/-3dB, it has pretty good bass, high-frequency response well beyond that of the CD, and does all that with reasonable consistency—within 3 decibels. Within 6 or more decibels is meaningless; within 1 or 2 is very good. Of course, notches or peaks in different parts of the frequency-response curve will result in different sound quality, and less scrupulous manufacturers will use that as an opportunity to lie like crazy with this spec.

It's the manufacturers who accurately specify **dispersion** who restore our faith in human nature. What they are doing is specifying frequency response as measured directly in front of the speaker, and then a few degrees off-axis, giving you both figures to suggest how the sound changes as you move out of the **sweet spot**. That's the acoustically perfect spot in which you are always supposed to sit and, in real life, cannot always sit. The more degrees off-axis a speaker can go, while maintaining uniform frequency response, the more consistent that speaker will sound as you move around the room.

Magnetically shielded models are designed to avoid picture distortion when placed next to a TV. The magnets in each speaker—something required in any conventional speaker, to make the drivers move—are constructed in multiples with an extra magnet to counteract the effects of the main magnet. This prevents them from generating magnetic fields that would turn your TV picture funny colors.

Finally, when buying any speakers for surround sound use, be sure to find out if **matching** center and surround speakers are available, and in what configurations. Part of the job of researching any speaker purchase is to investigate the entire line that interests you. Find out what the options are, what mates best with what, and look for matching driver sizes and materials.

Connections

- 5-way binding-post speaker terminals
- wire-clip terminals (spring-loaded or clamp-down)
- dual speaker terminals for biamping/biwiring

Most audio buffs agree that screw-down binding posts make better speaker terminals than wire clips. Depending on your definition of practicality, you might like the **5-way binding post** for its versatility. It can accept various kinds of terminated cable: **Spade lugs** are U-shaped connectors that fit around the post in a binding post. **Banana plugs** flex (or the good ones do) and fit into a hole in the post. **Dual banana plugs** are built into plastic casings that facilitate connection. **Pin connectors** are slim rods that fit into the hole in the side of a binding post or under the tab in a wire-clip terminal. Bare wire works with everything—to prevent oxidation over longterm use, you may solder the exposed copper, though that isn't strictly necessary to achieve a viable connection.

Some speakers use **wire clips**. The cheap version allows the user to simply lift a plastic tab and stick in the speaker-cable tip. There's also a sturdier cylindrical type becoming increasingly popular—it's all-metal and more durable. Wire clips on speakers are usually spring-loaded. They work best with bare wire though they'll also accept fancy cable terminated with slender pin connectors (which certainly do not sound better than bare wire, though they're more durable).

Some speakers have an extra pair of terminals to allow **biamplification** (separate amp channels feed each driver) or **biwiring** (separate cables feed each driver, joined at the amp). Biamping provides the most dramatic benefit, assuming the extra amp channels are available.

Shopping

Unfortunately, not everyone has local access to a top-flight retail listening room with the best acoustics and the broadest selection. If one's available, your local high-end emporium is a good place to start listening even if much of what it offers is beyond your budget. At least you'll get an idea of what good sound is like. And you might discover something wonderful that you'd never have found in a large chain store.

Here's an absolute must: *bring your own program material!* Familiar music will tell you more about a speaker than will a salesperson.

Listening conditions may contain snares. The most common sales ploy is to play two competing models at slightly different volumes. Because human ears hear more bass and treble at higher volumes, the speakers played louder will always sound better, even if the difference in volume is so slight as to leave the listener unaware of it. You can't be sure if your little receiver will perform as well as a high-end system's muscle amps and pristine pre-pros. There are many ways to be misled.

Give yourself a chance to hear how the bass response of a powered tower (or other floorstanding speaker) differs from that of a satellite & subwoofer set where the sub provides all the bass. Compare them both to a system based on bookshelf-sized speakers—looks dull, performs well, lots of manufacturers excel in this size range.

If you and/or the sales staff are more ambitious, ask if you may hear the same speakers—something you like—fed by a budget receiver, a middling one, a top-line model, and a separate configuration of pre-amp-processor and power amp. With or without subwoofer. Or with more than one subwoofer. If you'd really like to stir up trouble, ask: "Um, do speaker cables really make a difference?"

Surround receivers & components

A fancy TV will provide a wealth of switching possibilities but its built-in sound system is rarely strong enough to suspend disbelief. Basing a home theater system solely on a big TV is no more practical than building a mega-sound system to accompany a 13-inch bedroom set. A serious home theater sound system will be based either on a receiver (the best bet for most people) or on a combination of power amplifier and preamp-processor.

Receiver or separates?

- audio/video receiver
- preamp-processor
- multi-channel power amplifier

An **audio/video receiver** (or surround receiver) combines the functions of power amp, preamp (including volume control), surround processor, radio tuner, video processor, and audio/video switcher into one box. That saves money, space on the rack, and eliminates the mass of cables needed to keep an amp and pre-pro on speaking terms—it's the essence of practicality. The downside is that it's hard (though not impossible) for delicate surround processing circuits and beefy amp components to coexist in the same enclosure. When they share the same power supply, the amp tends to draw power away from the pre-pro at peak moments. But in a good design, it all works. Over the years manufacturers have conquered some pretty serious design challenges to make the a/v receiver a practical and even high-performing product.

77

There is, however, a higher-end alternative to receivers. Separating the **preamp-processor** from the **multi-channel power amplifier** allows both to operate at a more high-end level of performance, with more purity for the pre-pro, and more power for the amp. At least, that's the theory. In the real world, mating a cheap pre-pro with an inadequate amp can easily result in lower performance than a well-designed receiver could provide at the same price.

A high-end buff who knows the ropes can get the best performance from separate surround components, but the most practical reason to go for separates would be to fill a large room, something a muscle amp can do more easily than most receivers. If you love your old stereo power amp enough to keep it in your home theater system, a **three-channel amp** will provide the extra channels of amplification needed for the front-center and rear-surround channels. However, I don't recommend mixing amps. Some amps will react differently than others when you adjust the master-volume control, making it harder to keep all channels tracking up and down in a balanced manner.

If you're not sure whether separates are for you, buy a receiver with 5.1-channel analog line outputs. In a later upgrade, you'll be able to use those outputs to feed the inputs of a five-channel power amp. Still further down the upgrade path, you might consider replacing the receiver with a proper pre-pro.

Interpreting receiver/amp specs

- power rating (watts per channel)
- total harmonic distortion
- analog direct
- classes of amplification

The power rating for a typical receiver or power amp might read: "100 watts per channel, RMS, 20Hz-20kHz, into 8 Ohms at .05 percent THD." That means the amp delivers 100 watts to each speaker, as measured with a test tone, at full frequency response, with fairly low distortion, assuming that the speaker's nominal impedance is 8 Ohms. (See "Surround speakers/Interpreting speaker specs" for more about impedance.)

The Federal Trade Commission mandated this way of specifying power ratings decades ago following a scandal in which manufacturers were found to be playing fast and loose with their specs (and the truth).

To bring order to the chaos, the FTC decided power ratings would be measured using the **RMS** method: a continuous one-kilohertz test tone.

That gives today's power ratings a stable frame of reference—even though a one-note test tone really has nothing to do with music, which swells and falls and contains pitches ranging from low to high. For that reason, plus the human tendency to exaggerate, manufacturers used to specify **peak power**, measuring the amp's performance at a loud, transient high point. (With subwoofer amps, which are not separate products unto themselves and thus exempt from FTC rules, this still happens routinely. Take all non-RMS subwoofer amp specs with a grain of salt.)

FTC notwithstanding, manufacturers still cheat on ratings. One trick is to narrow the range of frequency response at the bass end to eliminate the demands of low bass reproduction. Another involves driving only one or two channels at a time, giving the power supply a break (an "all channels driven" spec is more valid). Yet another way to achieve a higher watts-per-channel spec is to specify an impedance lower than 8 Ohms (in other words, a speaker that accepts more current). As the impedance drops, the number of watts goes up. However, a manufacturer that specifies power ratings into 8 Ohms and at lower impedances (say,

Onkyo

Onkyo does an outstanding job of packing in the latest features at a competitive price. The TX-SR608 ($599) is a 7.2-channel THX Select2 Plus certified receiver with 3D-savvy HDMI 1.4a connectivity, Faroudja DCDi video processing, Dolby Pro Logic IIz height enhancement, and a full suite of Audyssey features including 2EQ auto setup and room correction, DSX height- and width-enhanced listening modes, and the one-two punch of Dynamic EQ and Dynamic Volume for greater comfort and clarity at low volumes—not to mention THX Loudness Plus.

79

6 Ohms) is providing useful information. Given side-by-side, the two specs will show you how the amp behaves when it feeds a more demanding speaker. Two different models may both deliver 100 watts into 8 Ohms, but if one delivers 120 watts into 6 Ohms while the other delivers 150 watts into 6 Ohms, the latter model delivers more current.

All other things being equal, more power is better. It makes your system play louder, sound better, and less likely to fry the speakers with harsh distortion. However significant the power rating may be, it would be a mistake to assume that amps with the same power ratings are identical. A good one is more likely to have a heavier power supply and higher current capability (check weight specs). To drive demanding low-impedance speakers, investigate outboard multi-channel amps.

Total harmonic distortion (**THD**) measures the amount by which an amp alters the waveform of a presumably pristine movie soundtrack or piece of music. It's generally assumed that distortion is harsh and nasty, although it can also sound mellow in vacuum-tube amps. Tiny differences in THD—say, .01 versus .05 percent—are rarely detectible by human ears. Some say differences of up to 1 percent are insignificant. Don't get caught up in the numbers game; it's the *nature* of distortion that determines whether, and how much, it's acceptable. Truly nasty distortion affects **higher-order harmonics**. (Harmonics are the higher tones that are sounded when a note is struck. They are multiples of the original tone. They're what enable you to distinguish middle C played on a guitar from middle C played on a piano.) The precise nature of distortion is something you won't find in the specs. Trust your ears.

If you have a high-quality analog signal source in your system, such as a turntable/phono-preamp combo, check whether a receiver has an **analog direct** (or **analog bypass** or **pure direct**) mode. Why? Because the average receiver translates all analog signals to digital. There's good digital and bad digital; you can usually assume that a receiver selling for a few hundred dollars represents the latter. Audiophiles rightly prefer to switch analog signals in the analog domain. Analog direct mode is most useful with full-range speakers. With sat/sub sets, which require digital bass management, analog direct is less useful.

Most surround receivers and power amps use **Class AB** analog amplification. Class AB is a hybrid of two other types of analog amplifier. **Class A**, occasionally used in high-end stereo systems, keeps both of its power stages full of current, which wastes a lot of power in the form of heat—though it does sound good. **Class B** pushes and pulls current between the two power stages, so that one is always on and the other off, which is more efficient, but not as sweet-sounding. Class AB keeps current in both stages part of the time but alternates between them for the remainder. This provides the ideal compromise between sound quality, efficiency, and cost.

Class D amplification is finding its way into a growing number of receivers and compact systems. These products convert the analog input signal to a train of pulses that is amplified by a rapidly switching output stage which is always either on or off. The amplified pulse train is low-pass filtered to recover the analog waveform and eliminate the ultrasonic switching noise. This process dissipates far less energy in the form of heat than conventional all-analog amplification and allows products to be made much smaller, slimmer, and lighter—and far more energy-efficient. Originally the sound quality of Class D lagged behind its elegant premise. But the technical hurdles are being overcome and some Class D products now sound excellent. Someday the majority of surround amps may be made this way. (Incidentally, the D in Class D does not stand for digital, and some manufacturers of Class D products object to them being described as digital amps.) Makers of Class D receivers include Pioneer and Rotel. One form of Class D technology, **ICE-power**, is licensed by various manufacturers from Bang & Olufsen.

Other alternative amplifier topologies include **Class G** and **Class H**. These analog amps shift from low-voltage rails to high-voltage rails in much the same way that a car shifts gears. This allows greater power output with same-size parts and/or greater energy efficiency. The difference between Class G and Class H lies in the rails—Class G uses multiple rails, while Class H uses variable rails—though the terms may also be used interchangeably. Arcam uses Class G in some products, while AudioControl uses Class H in some products.

Controlling your receiver

A receiver has many functions. So how you operate it affects what you'll get out of the product and how you'll feel about using it. There are several ways to control a receiver: through front-panel buttons and knobs, through an onscreen interface, or through a remote control. Some receivers also allow other means of control, such as an iPhone app or web browser interface.

Front-panel controls are usually rudimentary. Generally they'll let you power up the unit, select a source input, and select a listening mode. A few add extra functions behind a drop-down door, duplicating much

81

of what's on the remote control.

But for the most complete command of settings, including basic setup, you'll use the **graphic user interface**. It is almost invariably intricate—but it offers a good overview of what your receiver does and how it works. An above-average GUI makes adroit use of color and diagrams to clarify functions.

To operate the GUI and basic functions such as volume and mute from a seated position, you'll be at the mercy of the **remote control**. Does it have control codes for your other components **preprogrammed** into it? Is it capable of **learning** codes for other manufacturers' products? Look for backlit or glow-in-the-dark keys if you plan to use the remote in a traditional darkened home theater. And be warned that the simplest-looking remotes are not necessarily the easiest to use. See "Useful Accessories/Remote controls" for more pointers.

A few receivers support free **apps** that allow the **iPhone** or **iPod touch** to serve as the remote control. They must be certified "Works with iPhone" or "Works with iPod." If you've been using the portable device all day, you may like using it at night as well.

Some models also support a **PC and web interface**. The receiver has an IP address that can be entered into a PC's web browser. Once the browser recognizes the receiver, you can operate some functions. This is an especially good way to operate multi-zone features when you're not in the same room as the receiver.

Ease of use

Ease of use is a critical and often overlooked aspect of buying a receiver. This isn't just another set-top box or disc player—this is the heart of your system, and you're going to be using it a lot. Do the control menus make sense to you? On the remote, are there too many buttons? Or too few, making some oft-used functions less accessible? Are controls well differentiated by size, shape, color, labeling, or layout?

One thing that's making receivers much easier to use is **auto setup and calibration**. It places a small microphone—supplied with the receiver—in the listening position to gauge variables like speaker size, speaker distance, and acoustic conditions. The receiver then adjusts output levels for each channel and various other settings. If you haven't had much experience in setting up surround systems, auto setup will make your introduction to surround more painless and help you get the best performance out of your new system immediately. As time goes on you might become curious enough to buy a sound pressure level meter, measure the volume level coming out of each speaker, and fine-tune your receiver's settings manually. But for most beginners, auto setup is the way to go.

Another benefit of auto setup and calibration is that it corrects acoustic problems (see next section on "Room correction").

Room correction

Some high-end surround preamp-processors and receivers with automatic setup and calibration also have the ability to correct problems with room acoustics. This is a revolutionary development (and revolutionary is a not word I use lightly). Skillfully applied room correction can do a lot to improve performance, especially bass response.

Every room—except for the non-echoing chambers where manufacturers test speakers—has **resonant** or **room modes** that cause **peaks** and **nulls**. Even a speaker designed to provide a relatively flat frequency response under ideal testing conditions will sound louder at certain frequencies at home. Peaks that result from parallel walls are **axial modes**. Where four surfaces are involved, **tangential modes** result. **Oblique modes** involve the ceiling and floor as well as the walls. **Standing waves** are stationary patterns of high and low volume caused by waves bouncing between opposite walls. The frequency of the standing wave depends on the distance between walls. Effectively, the room becomes a resonator. Bloated midbass is a typical standing-wave problem.

These problems can worsen if certain room dimensions are identical or are multiples of one another. If two pairs of surfaces have a common resonance, the problem is even harder to correct. Wall construction can also contribute to resonance problems. For instance, a single layer of scantily reinforced sheetrock may resonate and muddy the room's sound considerably. That's why fancy multilayered sheetrock is becoming chic in high-end custom-installed home theaters.

Room correction measures room modes from one or more listening positions. Depending on the product, the correction may involve only one or two peaks, or inverse equalization may be applied across the whole frequency spectrum. Better room correction schemes work not just in the frequency domain, but in the time domain as well, adjusting the tendency of speaker drivers to operate out of phase. Getting them in phase results in a more coherent sound.

One way to apply room correction is to use a separate device called an **equalizer.** Equalizers can be **graphic** (adjusting fixed frequency bands that can't be changed) or **parametric** (with the ability to select the affected bands) or both. Setting them up often involves the use of simple microphones and a **real time analyzer.** EQs suitable for home theater use and real time analyzers are available from AudioControl and Gold Line, among others. THX-certified EQs are available from Rane.

But it's far more convenient to buy a surround receiver or preamp-processor that has room correction built in and performs it cleanly in the digital domain. Some products use technologies licensed from **Audyssey** or **Trinnov**, while others use their own proprietary technologies. For more about Audyssey, see "Understanding Surround Standards/Audyssey's auto-setup and listening modes." Most major manufacturers of surround receivers include auto setup and room correction. Pre-pros with room correction are available from Integra, Krell, Meridian, and TacT. Some Infinity, Mordaunt-Short, and Paradigm subwoofers also include room correction specifically for bass.

Room correction is not without its detractors. For one thing, they say, correcting for one listening position doesn't help others—a universal one-size-fits-all fix is not possible. That's why most room correction systems now measure more than one position to avoid favoring a single sweet spot. Poorly done room correction may also cause audible problems such as phase errors or ringing. On the other hand, minimal and carefully done room correction may help correct gross acoustic problems, especially with the interaction between subwoofers and room modes.

Video features of receivers and pre-pros

- video upconversion
- progressive-scan video processing
- lip-sync delay

Until recently the video-related features of surround receivers and preamp-processors were limited to video pass-through and switching. (You'll find more details about the ins and outs of video connections in a separate section below.)

But the latest receivers are getting slicker at the way they handle video. One helpful feature is **video upconversion**, the ability to convert between video formats so that signals entering the receiver through the lower-quality composite and S-video inputs can exit the receiver, en route to the video display, through the higher-quality HDMI or component video outputs. That doesn't improve the quality of the signals, but does make your system easier to set up, because you need to connect only the highest-quality video output to the display.

Another prominent video feature is **1080p processing**, which line-doubles 1080i signals to 1080p, and converts others to 1080p as well. Similarly, **progressive-scan video processing** converts the half-frames of an interlaced (480i) signal to the full frames of a progressive-scan (480p) format, smoothing out jagged edges and other video artifacts. Some receivers have extremely high-quality video processors licensed from the likes of Anchor Bay and Faroudja.

Video processing circuits (in the display and disc player as well as the receiver) can cause a lip-sync problem onscreen because they take so much time to swallow up and manipulate whole frames of video. When the video lags the audio by a few seconds, this causes a disconnect between the voices you hear coming out of the speakers and the lips you see moving onscreen. An average person can detect as little as a single film frame's worth of delay—people who transfer films to video can detect as little as a half-frame. Many receivers offer **lip-sync delay** to delay the audio so that the video can catch up.

Network audio features of receivers

Any receiver can accommodate orthodox source components such as disc players and set-top boxes. Most now accept **iPods** as well, either directly via USB or more often through an accessory dock. Some receivers also accept a **Bluetooth** adapter, allowing access to music from a cellphone or other portable device.

But step-up models go a step further by including attractive network features. Connect the receiver's ethernet jack to your broadband router and you can get access to one or more of the following: media from your PC's hard drive, **internet radio**, or a subscription music service. Internet radio can be through a generic interface or through a special service such as **SHOUTcast** or **Pandora**, which makes it easier to find and select stations. Subscription music services include **Rhapsody** and **Napster**.

Some receivers with network features are certified by the **Digital Living Network Alliance**. **DLNA**-certified receivers can communicate with a PC's hard drive through the router and the Windows Media

Player (if you enable file and media sharing in WMP) to access music, video, or photos. Other products with DLNA certification include TVs, Blu-ray players, PCs, cameras, and mobile devices. The latest version of DLNA is 1.5, circa 2006.

Introduction to receiver connections

"There comes a time in the affairs of man," W.C. Fields once said, "when he must take the bull by the tail and face the situation." Yes, it's time to think about what jacks you need on your receiver's back panel. Count your components. Then consider whether the receiver has enough inputs and outputs of the right kind to suit your system's current and future needs. (For the sake of simplicity, this section will refer only to receivers, but the information applies to pre-pros as well. You may also refer to the "Connection glossary" under "Connecting a Home Theater System" and to chapters on other products.)

Highest-quality video and audio connections

- HDMI digital interface
- IEEE 1394 digital interface
- component video inputs/outputs
- 5.1- to 7.1-channel analog line inputs and outputs

HDMI and IEEE 1394 are two highest-quality digital interfaces on receiver back panels. HDMI is by far the dominant one. It is contributing to the long-desired uncluttering of receiver back panels. When these digital connections are not available, the next best options are their analog equivalents, component video and 5.1- to 7.1-channel analog audio.

The most desirable jack on the back of a receiver is **HDMI**. It carries both video and surround (or sometimes just stereo) audio. It's available in multiple versions from 1.0 to 1.4a. Don't assume that the latest version is supported in every receiver. Though we covered this in the chapter on "DTV connections," here's the rundown again: The original HDMI 1.0, released in 2002, supports video and stereo audio, not surround. HDMI 1.1 adds Dolby TrueHD, Dolby Digital Plus, DTS-HD High Resolution Audio—the new cutting-edge lossless surround standards that improve the audio performance of the nascent Blu-ray video-disc format—and DVD-Audio, a high-resolution music format. HDMI 1.2 adds Super Audio CD, the main rival to DVD-Audio, and is more computer-friendly. HDMI 1.3 came out in 2006 and supports higher bandwidth, greater color depth, a new mini-connector for camcorders, and automatic lip-sync to keep soundtracks coordinated with faces speaking onscreen. It also supports DTS-HD Master Audio. HDMI 1.3a and 1.3b add minor testing refinements not relevant to the consumer. HDMI 1.4 has rudimentary 3D compatibility, and lets devices share internet connections and data, among other features. HDMI 1.4a has greater 3D compatibility. Get 1.3 or higher for the best sound or 1.4a for the most up-to-date 3D video.

A note on connecting an SACD player to a receiver: The best method is a direct DSD signal via HDMI. The second best method is to set the player to output high-res PCM via HDMI. The third best method is the multichannel analog interface, bypassing HDMI.

The current use for **IEEE 1394**, mainly on Denon and Pioneer products, is for high-resolution audio signals from DVD-Audio and SACD players. However, 1394 is also capable of carrying video and for that reason appears in camcorders.

Component video inputs/outputs are next in the videophile pecking order. Like HDMI, they're HD-capable, albeit analog. Component connections are bulky—using a trio of red, green, and blue RCA-type plugs—and inadequate bandwidth can kill their advantages. According to the vigilant videophiles of the Imaging Science Foundation, few receivers or preamp-processors provide adequate bandwidth for component video connections, so you might be better off making component video connections directly between your DTV and source components.

Receivers with **multichannel analog line inputs** can be fed by disc players with **multichannel analog line outputs**. This arrangement came about in the early days of Dolby Digital, when there were still many receivers that didn't support it. Nowadays those multichannel-ins are still useful, but for other things. Possibilities include players handling Dolby TrueHD, Dolby Digital Plus, DTS-HD, a DVD-Audio/SACD universal player, and future surround formats. Because several new surround formats are 7.1-channel, multichannel analog line outputs on many products add the extra pair of back-surrounds to the basic 5.1-channel array. When a disc player can convert high-res audio signals to PCM via HDMI, that is a better form of connection than multichannel analog.

Among the receiver's 7.1-outs is at least one **subwoofer output**, called **LFE** (low frequency effects). It is usually a single RCA-type jack. The sub may have stereo inputs, but feeding just one of them is fine—

the sub doesn't need a stereo signal. (The other method of connecting a sub is to go from the front left/right speaker outputs, through the sub's speaker-level ins and outs, to the speakers. This will make use of the sub's **high pass** filter—in other words, the sub will strip out the low sounds for its own use and send the rest of the signal to the main speakers. A few subs support high-pass filtering via line-level connections.)

A receiver with **5.1- to 7.1-channel analog line outputs** can serve as a surround preamp-processor when plugged into a multi-channel power amp.

Other video connections

- S-video inputs/outputs
- composite video inputs/outputs
- front-panel audio/video inputs
- DVR/VCR inputs/outputs
- monitor outputs

After component video (above), the round multi-pin **S-video** is the second best analog input/output. Unlike composite video (below) it separates the brightness and color parts of the signal to reduce video artifacts such as dot crawl and moiré. Some receivers now omit S-video.

All audio/video receivers have **composite** video inputs and outputs color-coded with a yellow RCA plug. Nearly all products that provide any kind of video connection have composite video. It is inferior to S-video and component video because it mixes the brightness and color portions of the signal, leading to buzzing rainbow-like moiré distortion.

Front-panel audio/video inputs are typically analog stereo audio with either composite or S-video. An increasing number of receivers are offering front-panel HDMI, which is preferable. Front inputs make it easy to plug in a videogame, camcorder, or other source component.

DVR or VCR inputs/outputs mate the audio ins/outs with composite video (always) and S-video (sometimes).

Monitor outputs let the receiver send a video signal to the TV. Be sure any receiver you buy has a monitor out that matches the highest-quality input of your video display. Today that usually means HDMI.

Other audio connections

- digital audio inputs/outputs (optical or coaxial)
- stereo analog line inputs
- multi-room/multi-zone, multi-source
- headphone output

Most receivers provide **digital audio** connections, mostly inputs, sometimes accompanied by one or two outputs. These digital connections may be labeled **PCM** (which stands for **pulse code modulation**, an uncompressed digital signal) or **S/PDIF** (Sony/Philips digital interface).

There are two types of digital jacks, **optical** (also known as **Toslink**) or **coaxial**. (The term coaxial wins the confusion sweepstakes. Strictly speaking, it refers to a type of cable construction with an outer sheath surrounding an inner core. More generally, it refers to a variety of cables with different-sized plugs and different uses, from digital cables to the RF cables used by cable TV systems.) Optical cables offer the advantage of defeating **hum**—an audible and unwanted low tone coming out of your speakers—because they pass the signal optically rather than electrically, thereby breaking up the closed circuits known as **ground loops**. Even so, some prefer the coaxial type (with RCA connectors) for its allegedly superior sound quality. Coaxial cables are sturdier than optical cables, which strangle the signal when bent. And the coaxial connection allows you to substitute inexpensive composite video cables for expensive digital audio cables, as long as the yellow-color-coded cables have a characteristic impedance of 75 Ohms.

Except for Blu-ray's lossless surround and SACD, which are best delivered by HDMI or the multichannel analog interface, digital inputs are the best way to connect a DVD or CD player to your receiver, though there are alternatives. You can also connect a disc player with 5.1- to 7.1-channel analog line outputs to a receiver or pre-pro with 5.1- to 7.1-channel analog line inputs. If you'd prefer to listen to Blu-ray discs or DVDs in stereo, you can use the player's stereo (or **downmix**) outputs, and select the two-channel option from disc and player menus. Those digital jacks are also good for plugging in CD and MiniDisc equipment. If you do digital recording with a CD or MD machine, it can be helpful for the receiver to have digital out as well as digital in.

Stereo analog line inputs accept CD players, cassette decks, and any device with analog outputs. If you still want to record cassettes,

89

make sure there's a stereo analog line output as well, possibly labeled "tape loop" (see below). Most receivers provide more of these than you'll need, but if you've got a rack full of legacy formats, make sure you have enough.

Most receivers convert signals entering through their analog inputs to digital. If you're plugging in a CD player or other digital device, the signal must go through the CD player's digital-to-analog converter, then through the receiver's analog-to-digital converter. In this situation, use a digital connection between CD player and receiver to bypass those unnecessary D/A and A/D stages—you'll get cleaner sound.

Sometimes a receiver with an analog direct mode will sound better with any device that has a good-sounding analog output (including phono preamps and bleeding-edge CD players).

Higher-end receivers and pre-pros have **multi-room** (or **multi-zone**) **outputs**. These may be just for audio, or for both video and audio. The video connections are generally low-res composite video. The audio connections are usually stereo, and can be analog, digital, or speaker outputs. **Multi-room/multi-source outputs** can process two different sound sources for two rooms. One family member could sit in the livingroom watching a DVD while another sits in a nearby bedroom listening to a CD.

A receiver's **headphone output** will usually be the **1/4-inch** type that accepts what used to be called a phone plug (visualize an early 20th century Bell System operator jamming those big plugs into a patch bay). Portable devices are more likely to use the **1/8-inch** type, also known as the **mini-plug/jack**. If you want to use mini-plug headphones with your receiver, RadioShack or any electronics supply store will have the simple 1/8- to 1/4-inch adapter you need.

Portable player connections

- iPod dock
- USB, WMA

The **iPod dock** is a natural extension of the receiver. Using the 30-pin docking connector on most iPods, it lets the player send audio signals into a big system. Some docking stations also handle video. They usually recharge the player (an alternative to the PC-USB link or extra-cost AC adapters). The player's screen may be disabled—but depending on the design, you can use either an onscreen display through your video monitor, front-panel receiver display, or both. An advantage is the ability to operate the iPod with the receiver's remote—since the iPod doesn't come with one. Some receivers accept iPods via direct USB input, so you can plug the white iPod cable directly into the receiver.

A few receivers include a **USB** jack for players outside the iPod universe as well as external hard drives or thumb drives. Some receivers support Microsoft's **WMA** (Windows Media Audio) file format via USB.

Radio connections

- AM/FM antenna inputs
- HD Radio
- XM or Sirius satellite radio
- ethernet

If you listen a lot to the radio, your receiver may well have **AM/FM** antenna inputs, but they lead to a barebones chip-based tuner that won't be sensitive enough to pull in far-off stations. You might either add an outboard AM/FM tuner to your system or graduate to surround separates including a good tuner/preamp. Anything lacking AM/FM and phono capability is classified as an **integrated amplifier** or something other than a receiver.

HD Radio, a digital supplement to AM and FM over-the-air broadcasting, is supported by a small but increasing number of receiver makers. It allows multiple channels in a single broadcast frequency.

A few surround receivers support **satellite radio**, either XM, Sirius, or both. You'll need a satellite antenna, typically sold as a $20 add-on.

Ethernet connections are just starting to appear in surround receivers for internet radio and PC link. They accept a home network connection from a router, with Cat5 wiring, or from a wireless bridge.

Legacy connections

- phono input (moving-magnet or moving-coil)
- ground terminal
- tape-loop inputs/outputs (for analog recording media)
- RF-modulated Dolby Digital input (for laserdisc)

Still got a shelf or two of LPs? No serious music lover or audiophile would make fun of you for that. But your turntable will need a **phono**

input. It's not a line-level input such as the analog jacks that talk to your cassette deck. A phono input accepts the fragile sub-line-level signals generated by tiny magnets and coils within your phono cartridge when its stylus wiggles in those mysterious black grooves. Unless indicated otherwise, most phono inputs are compatible with **moving-magnet** phono cartridges, which are relatively reliable, don't cost much, and produce good bass. However, some receivers' phono inputs are switchable for either moving magnet or **moving-coil** cartridges, which according to audiophiles can more accurately reproduce high frequencies. Most receivers with phono inputs also provide a **ground terminal** to accept another connection that defeats hum in the turntable/cartridge. If your receiver has no phono input, or you want higher-quality vinyl playback, you'll have to dedicate one of its line-level inputs to a separate **phono preamp** (which might accept one or both types of cartridge). In addition to line-level outputs, some phono preamps also have USB outputs, useful for ripping vinyl, and/or headphone outputs.

Analog **tape loop** jacks include both inputs and outputs. That not only allows you to record and play without swapping cables—it also helps you monitor recordings in progress when using cassette decks and other analog recording media. Some preamp-processors provide **record and listen** controls that allow you to record from one input while watching/listening to another.

Out of respect for one of home video's most distinguished dead formats, some manufacturers add an **RF-modulated digital input** to accept surround signals from the small number of laserdiscs (LDs) encoded in Dolby Digital.

Control connections

- IR remote jacks
- 12-volt trigger
- RS-232

Every receiver has an IR (infrared) window on the front panel to accept remote control signals. However, if the receiver lives in a cabinet, without a direct line of sight from seating area to front panel, remote control signals can't reach the window. Therefore some receivers have one or more **IR remote jacks** on their back panels to connect IR repeaters that can accept remote commands from the seating area.

To coordinate the powering up and down of components, such as multiple amplifiers in a large-scale system, you need a **12-volt trigger**—though some versions of HDMI also coordinate components.

RS-232, from the computer sphere, allows firmware upgrades, a touchscreen interface, and other custom-install operations.

Speaker connections

- collared binding-post speaker terminals (UL-approved)
- wire-clip terminals (spring-loaded or clamp-down)
- A/B speaker terminals (binding post or wire clip)

As stated in the chapter on speakers, binding posts are superior to wire clips. In receivers the most common type is the **collared binding post**. It's less versatile than the five-way binding post, but does a better job of concealing wire tips from curious little fingers, reducing the likelihood of electrocuting an unwary child or pet. At least, that's what the Underwriters Laboratories believe, and for legal reasons many receiver makers (and some high-end power-amp makers) follow suit. Collared binding posts make it impossible to use spade lugs, which audiophiles prize, though they do allow banana plugs and bare wire. Binding posts provide a more secure and better-sounding connection than wire clips.

Occasionally receivers substitute **wire clips** for binding posts, at least for some channels. There are two kinds of wire clips. Though either one is perfectly capable of snapping to pieces in your hand, the **spring loaded** type is marginally more durable, and usually allows use of thicker (16-gauge or 12-gauge) wire tips. **Clamp down** wire clips are found on very low-end receivers and inexpensive compact systems and are to be treated with the same delicacy as Faberge eggs. In terms of child safety, a wire clip is as hazardous as any binding post, because it leaves bare wires exposed.

Many receivers provide two sets of **speaker outputs** labeled **A** and **B** for the front left and right channels. The B set parallels the A set and allows connection of a second set of stereo speakers.

Most receivers come with seven channels of amplification. The last two channels can be configured for a variety of uses including back-surround, height, or width speakers; or for biamplification of the front left and right speakers; or for second-zone speakers.

Shopping

Contrary to myth, receivers with the same power rating do not all sound alike. Even if you've found a few models with the right power, connections, and other features, sound quality still is not a settled issue. As with any audio shopping expedition, bring your favorite listening material—Blu-ray discs, DVDs, CDs, LPs, whatever—and be prepared to listen your ears off.

If possible, audition receivers with your own speaker models. If that's not possible, find out what you can about the speakers being used for the audition—their enclosures, driver sizes, and power handling ratings. It never hurts to know what you're listening to. Of particular importance are sensitivity, which governs how loud different speakers play when given the same amount of power, and impedance, which must be carefully matched when mating speakers and receivers or power amps.

The first thing to determine is how much volume the receiver can support with the connected speakers (assuming they're similar to yours). It's not unusual for a receiver to use half of its volume range to play loudly—say, 85 decibels, which you can measure with a cheap sound meter. Even two-thirds may be acceptable. Beyond that point the receiver is straining—or **clipping**—as it runs out of power. Then you should move up to something more powerful.

Cycle through all the receiver's modes. Listen to it in DTS-HD Master Audio—that's how most of your Blu-ray discs will sound. Try Dolby Digital—that's how your DVD collection will sound. Listen to it in Dolby Pro-Logic II—that's how your old VHS tapes and CDs will sound. Is it too aggressive, harsh? Too vague, mushy? It's OK for a receiver lean to one side or the other, but if it does not perform well in the modes you use most often, best consider another model.

Next, switch to stereo mode, leaving only the front left and right speakers operating. Crank up the volume. If you prefer not to use surround modes when listening to music, this is how your stereo CDs, LPs, and cassettes will sound. Even if you prefer listening to music in surround, stereo is a less forgiving medium than surround. That makes it a good test, and will suggest how you'll later feel about the receiver's overall performance.

By now you've spent some time with that remote in your hand. Love it? Hate it? Are the control menus easy to navigate? If not, move on to more promising models.

Never buy a receiver as an impulse decision.

94

Understanding surround standards

Surround sound literally gets more complicated with each passing year. Among several organizations vying for your surround-sound dollar are **Dolby Laboratories**, **DTS**, **THX**, and **Audyssey**. These companies do not make hardware. Instead, they license their surround technologies to other companies that do make hardware. Their conflicting *and* overlapping formats form a thicket that turns on a few droolers and turns off the rest of us.

The upside is that you can choose to have a system with between 5.1 and 11.1 channels. Software encoded in 5.1 sounds fine on 7.1 systems, and by the same token, 7.1 software won't trip up a 5.1 system. Diversity means more consumer choice.

The downside of surround format proliferation is that it frustrates the universal consumer desires for clear, stable standards and general ease of use. If all but the Dolby formats disappeared, home theater buffs would still have access to jaw-dropping surround sound in a variety of delivery media. In the longterm, if enough people are turned off by the riot of surround formats, the confusion may even slow consumer acceptance of surround sound in general. The ultimate winner of this increasingly crowded race may be our old friend stereo.

Very important note

You'll probably find the following discussion of surround standards incredibly complex and confusing. However, the most significant ones fall into just three categories: current generation, previous generation, and next generation. In other words: current digital formats using 5.1+ channels, older formats that convert stereo to 5.1+, and brand-new next-generation formats that provide the highest quality. If you look at the bullet points in the introductory sections immediately below, you'll see all Dolby's and DTS's nomenclature sorted by generation. Our tour will meander in the same direction—current, old, and next-gen. See you at the end, and please forgive me.

Introduction to Dolby standards

- Dolby Digital 5.1, Dolby Digital EX
- Dolby Pro-Logic, Dolby Pro Logic II/IIx/IIz
- Dolby Digital Plus, Dolby TrueHD

Whether your source of surround-encoded programming is Blu-ray, DVD, VHS, satellite, cable, or broadcast television, Dolby encoding formats cover all these bases and more. While cineplexes use a variety of surround formats, in home theater systems the dominant player is **Dolby Digital 5.1,** available via BDs, DVDs, DTV broadcasts, satellite, and cable. Dolby Digital is a 5.1-channel format. Its 6.1-channel variant is **Dolby Digital EX**. **Dolby Pro-Logic**, **Pro Logic II**, **Pro Logic IIx,** and **Pro Logic IIz** decode the oldest form of analog surround, adapt stereo sources to surround, and add extra channels to surround sources. The newest members of the Dolby family are the more efficient and higher-performing **Dolby Digital Plus** and the no-holds-barred, cutting-edge **Dolby TrueHD**. All of these are discussed in later sections.

To learn more about the Dolby surround standards, visit dolby.com.

Introduction to DTS: the un-Dolby

- DTS 5.1, DTS-ES Matrix, DTS-ES Discrete, DTS 96/24
- DTS Neo:6
- DTS-HD High Resolution Audio, DTS-HD Master Audio

Although the Dolby-licensed surround formats dominate home theater systems, they do have a competitor in **DTS Inc.**, originally **Digital Theater Systems**. Like the Dolby formats, DTS originated as a system for moviehouses, where there is more diversity in surround formats (also including, for example, **SDDS**, or Sony Dynamic Digital Sound). And like Dolby, DTS has steadily infiltrated home surround gear.

The family of DTS standards includes: **DTS 5.1**, a competitor of Dolby Digital; **DTS-ES**, a competitor of Dolby Digital EX; **DTS 96/24**, a lossy but high-resolution format used in some DVD-Audio titles; and **DTS Neo:6**, a six-channel competitor of Dolby Pro Logic II and IIx. In addition, DTS has recently brought out two new high-resolution standards: **DTS-HD Master Audio** and **DTS-HD High Resolution Audio**, the latter dominant in Blu-ray.

To learn more about DTS, visit dts.com.

Dolby Digital 5.1 and DTS 5.1

These two current-generation surround standards dominate DVD and other digital program sources. They have 5.1 channels. Their 6.1-channel extensions will be discussed in the next section.

As originally designed by Dolby Laboratories, Dolby Digital is a 5.1-channel surround format that delivers three **discrete** (meaning separate) channels of information to the front left, center, and right speakers and two discrete channels of information to the surround left and right speakers. Dolby Digital also provides a discrete subwoofer channel labeled **LFE** (**low frequency effects**). That makes it an improvement over previous Dolby surround formats which used **matrixing** to derive the front-center, surround, and sub channels from front left and right.

To efficiently provide all this audio information, Dolby Digital uses **perceptual coding** to eliminate audio data that is masked by louder sounds and therefore unheard by the human ear. This so-called **lossy** encoding is not an audio meat grinder—it is based on sophisticated psychoacoustic principles and can provide surround sound of great subtlety and power. Dolby Digital usually provides 5.1 or 6.1 channels at 384-448 kilobits per second, quite a feat compared to, say, the MP3 format, which needs at least 192kbps just to make stereo sound adequate. The standard's full range is 96-640kbps. Any Dolby Digital-compatible surround receiver can accept up to the top limit. DVD is limited to 384-448kbps, though Blu-ray's top limit is 640kbps.

Dolby Digital performs other kinds of signal processing. Some, such as **dialogue normalization**, work automatically in the background. Dialogue normalization makes adjustments to overall levels to smooth out differences between program sources, keeping the words of performers intelligible, among other benefits. **Dynamic range compression**, also known as **midnight movie mode**, is switchable and under the viewer's control. This function reduces the disparity between the louder and softer portions of the soundtrack, using volume-level codes embedded in Dolby Digital soundtracks by the mixing engineer, while ensuring that the soundtrack retains some of its original dynamic range.

Next we move from Dolby Digital 5.1 to **DTS 5.1**, also known as **DTS Digital Surround** or just plain **DTS**. For nearly everything Dolby offers in the consumer sphere, there's a competing flavor of DTS, and it's important to understand the rival standards in context.

97

DTS 5.1 appears as an alternate soundtrack on some DVD-Video releases, in addition to Dolby Digital, and occasionally it pops up as an alternate soundtrack on DVD-Audio releases, instead of Dolby Digital. Most DTS soundtracks have 5.1 channels. (We'll get to the 6.1-channel versions of DTS and Dolby Digital in the next section.)

DTS purports to provide a higher-quality method of surround encoding. Like everything from Dolby Digital to MP3, DTS uses perceptual coding to eliminate audio data that are inaudible to human ears—in other words, it is classified as a lossy format. DTS is usually encoded at a higher **data rate** (the number of bits per second) than Dolby Digital. The range is 768 kilobits per second to 1.5 megabits per second, versus a maximum of 640kbps (and in practice, more often 384-448kbps) for Dolby Digital. Some say this produces less edgy sound, but that also makes it less efficient, effectively reducing the number of bits that can be allocated to video on a DVD, and thus reducing the quality of video. Both systems have variable data rates, allowing mastering engineers to decide how much space on a DVD is devoted to sound and how much to picture.

Whether DTS-encoded DVDs are superior to those with mainstream Dolby Digital is hotly debated by home theater buffs, not to mention the Dolby Labs and DTS people. It would be fair to say that Dolby Digital and Dolby Surround encoding are far better entrenched in DVD movie releases as well as broadcast, cable, and satellite television. Most movies on videotape with stereo soundtracks are also encoded in Dolby Surround.

Dolby Digital EX and DTS-ES

These current-generation surround formats add extra channels and speakers to the standard 5.1-channel array. Referred to as **6.1-** or (incorrectly) **7.1-channel** formats, they originated in movie theaters where there's a genuine problem in achieving even surround coverage for the folks in the back rows. However, is your home theater the size of a real moviehouse? Does it seat hundreds? Didn't think so.

Dolby Digital EX was developed by Dolby Labs and THX. Originally licensed (with great success) through THX as **THX Surround EX**, it is now also licensed through Dolby as Dolby Digital EX, while THX has proceeded to make EX a feature of its revamped **THX Ultra2** standard.

In any event, EX adds one extra channel, and one or two extra speakers, in what has become known as the **back-surround** position (though I find the term **rear center surround** more precise). While the regular surround (or in this context, **side-surround**) speakers are placed *toward* the back of the room, along the side walls, the back-surround speakers go against the back wall. The two back-surround speakers receive the same signal. Technically, that makes Dolby Digital EX a 6.1-channel system—if you count data channels, not amp channels or speakers—though you will often see it referred to as a 7.1-channel system.

Recognizing that not every home theater system can accommodate an extra pair of back-surround speakers, some surround processors include a **phantom back center** mode that accepts the EX signal and simulates back-surround effects using the two side-surround speakers. This is a viable option to tearing up an existing 5.1-channel system.

In any event, that back-surround signal is derived from the surround left and right channels using the same **matrixing** process that Dolby Pro-Logic uses to derive front-center and surround signals from the front left and right. Doing away with matrixing is one of the key improvements that Dolby Digital has made over Dolby Pro-Logic. Thus, in attempting to take a step forward, EX takes a step backward, emphasizing the quantity of surround channels over quality.

The 6.1-channel version of DTS is **DTS-ES Discrete**. It encodes the back-surround as a fully separate and independent channel. That makes it superior to EX which leans on the old trick of matrixing the sixth channel from the surround left and right channels. For compatibility with non-discrete DTS-ES programming, DTS also provides a matrixed version of the back-surround channel (**DTS-ES Matrix**).

Dolby Pro-Logic and Pro Logic II

Having discussed the chief current-generation surround formats, let's take a step backward to those that exploit an older form of surround encoding. Their second—and important—function is to adapt stereo sources, like CDs, to surround.

Surround sound in the home started with **Dolby Surround**, a few years after it made a big theatrical splash in *Star Wars*. Dolby Surround's great strength is pure efficiency. It can travel in two channels, which various surround modes will expand to both front and rear.

Though Dolby Surround signals are still around—in VHS tapes and analog cable channels, among other places—it didn't last long in home surround gear. It was replaced by **Dolby Pro-Logic**, which uses the same Dolby Surround signal, but decodes it into four channels: front left, center, right, and a monaural surround channel served by two speakers, plus subwoofer. The addition of the front-center channel makes dialogue more intelligible and gives the soundfield some much-needed tightening. And of course, the subwoofer channel enables a system to produce the kind of monster bass home theater buffs have come to crave.

Enabling this neat trick is a process called **matrixing** which buries sonic information for the front-center and rear-surround channels in the front left and right channels of a stereo soundtrack. Dolby Pro-Logic has a forgivingly soft-focused sound that lets orchestral soundtracks breathe and makes dialogue and sound effects a little less harsh—in other words, sounds good, turn it up! Pro-Logic's downside: It's a little less detailed, especially in the rear-surround speakers, which it serves with just a single monaural signal—and a truncated one at that, cut off at 7kHz, well within audible range.

Pro-Logic never quite gels with CDs and other stereo music sources, and that's a problem. You can't just leave your receiver in Pro-Logic mode all the time and expect stereo music to sound as good as surround-encoded movies. Given stereo source material, Pro-Logic has a nasty way of shoving too much of the mix into the front-center channel. The end result sounds more like mono than surround. Mindful of its flaws, surround design guru Jim Fosgate set out to achieve a cleaner sound that meshes effortlessly with any stereo source, and licensed the result to Dolby Labs.

There's more to the resulting **Dolby Pro Logic II** than the addition of a roman numeral. Dolby Pro Logic II does the following differently than the original Pro-Logic: Breaks the mono surround channel into a pair of stereo surround channels. Provides more sophisticated filtering (or none at all) for surround effects. Provides user control of the center image and front/rear balance. Does all this with simpler processing, resulting in cleaner sound. And, in the most radical departure, is designed for use with stereo as well as surround-encoded source material.

Pro Logic II has four modes: **music, movie, game**, and **emulation**. The music mode has a more spacious feel than the dialogue-oriented movie mode. Music mode sounds superb—you might prefer it over stereo for two-channel sources. Emulation is merely a reproduction of the original Pro-Logic. It is inferior to the new music and movie modes and can be safely disregarded.

100

Pro Logic II's **dimension control** is an especially powerful tool (and would be even more powerful if user interfaces allowed easier access to it). It works like a front/rear balance control. Two kinds of material become easier to listen to with this control at your fingertips. The first is the stereo-turned-surround mix with so much "phasey" (hollow, disembodied) material that it drives the decoder into a rear-aggressive sound. You could move much of the rear-surround material toward the front soundstage, so you feel more a part of the audience, relieving any discomfort at being too near the performers. The second is the flat, clean mix with everything miked up close and very little room ambience. You could break up the tightly focused front soundstage to let it breathe toward the rear.

The **panorama control** further heightens surround effects by leaking some information from front to rear. This deepens the soundstage, and may benefit some recordings, though for maximum realism you might prefer not to use this option.

The **center width control** distributes information among the three front channels. It can either push things inward toward the center channel, or push them outward toward the left and right channels.

There is no 5.1-channel DTS equivalent of Dolby Pro Logic II, but there is a DTS Neo:6 (see next section).

Dolby Pro Logic IIx, Dolby Pro Logic IIz, and DTS Neo:6

To accommodate evolution toward 6.1- and 7.1-channel surround systems, Dolby Pro Logic has received a further update with **Dolby Pro Logic IIx**. This format expands two-channel as well as 5.1-channel source material to 6.1 or 7.1 channels, deriving the one or two added channels from the existing ones. Think of it as a retrofit for sources that have fewer than 6.1 or 7.1 channels—its purpose is to fill speakers in an expanded surround system that otherwise would lie silent. DPLIIx affects a wide range of two-channel software including videocassettes with Dolby Surround encoding, CDs, MP3s, and audiocassettes. It also works on 5.1-channel sources such as Blu-ray and DVD titles and HDTV broadcasts with Dolby Digital 5.1, DVDs with DTS 5.1, DVD-Audio, and SACD. Like Pro Logic II (minus the x) DPLIIx has separate modes for music and movie playback and adds a third mode for videogames.

Why did Dolby name this new format Pro Logic IIx rather than Pro Logic III? I can only speculate—and here goes. The name may reflect an ambivalence about undermining the 5.1-channel surround standard that has done so much to popularize home theater. Dolby's own DPLIIx FAQ states in part: "Dolby strongly believes in the intrinsic value and benefit of the 5.1 surround platform.... The introduction of 7.1 playback by Dolby should in no manner be interpreted as an abandonment of the 5.1 playback medium, or of its obsolescence as a format." DPLIIx is simply an adaptation to existing market trends. But don't let the hyping of new surround formats goad you into tearing up a perfectly good 5.1-channel system. Any existing investment in 5.1-channel surround should provide pleasure for years to come.

The newest member of the Pro Logic family is **Dolby Pro Logic IIz**. It does everything IIx does and adds two **height channels** above the main front left and right channels. DPLIIz, explains Dolby, "processes low-level, uncorrelated information—such as ambience and some amorphous effects like rain or wind—and directs it to the front height speakers." This makes the front of the soundfield a little airier though the effect is usually subtle—I spent my first day with DPLIIz rushing to the left height speaker and holding it to my ear. With the right movie material, it can be occasionally transformative, but not with all movies. With music, the effect is benign but not a game changer. DPLIIz is not encoded into movies at present though it is encoded into a few games. Bottom line, surround is still a 5.1-channel medium at heart. I wouldn't recommend tearing up an existing system to add height.

The DTS counterpart of Dolby Pro Logic II and IIx is **DTS Neo:6**. It has a slightly "wetter" sound than DPLII, with more reverb in the surround channels, and less closely approximates the original feel of stereo mixes than Dolby Pro Logic II.

Audyssey's counterpart to Dolby Pro Logic IIz is **Audyssey DSX**, which adds height as well as width channels. More on that later.

The latest from Dolby: Dolby Digital Plus, Dolby TrueHD

Our tour of all things surround has passed through the current and older generations. Now we move into the next generation of surround standards. Aren't you excited?

Dolby Digital Plus first found favor in European satellite delivery. In the United States, it made its first impact in the dueling high-def DVD standards, the surviving Blu-ray and now deceased HD DVD. It is also approved for use in ATSC, the standard for over-the-air DTV broadcasting in the U.S., as a robust channel (with error correction).

Dolby Digital Plus is not so much a replacement of the prevalent

Dolby Digital standard as an extension. It supports up to 14 channels. The data rate can run from 96 kilobits per second to 6 megabits per second, and because it is more efficient, it works in satellite video, streaming media, and other low-bandwidth applications.

But doing the same with less is only part of Dolby Digital Plus's attraction. It can also do more with more, improving sound quality by increasing the data rate from Dolby Digital's average of 384-448 kilobits per second to as much as 6 megabits per second. That huge increase in data should provide a larger, warmer soundfield with more vocal clarity while virtually eliminating any audible effects of compression.

Dolby Digital Plus may help Dolby Labs leapfrog its competitor DTS by supporting full discrete channel separation in 7.1, 6.1, or 5.1 surround formats, something DTS-ES can do (in 6.1) but Dolby Digital EX cannot do (in 6.1 or 7.1). Mixes are totally independent—so if you prefer 5.1 channels, you'll get a dedicated 5.1-channel mix, with all channels in balance, not 7.1 stripped down to 5.1. Plus is also better at handling high frequencies, transient elements, and continuous tones than regular Dolby Digital. To maintain compatibility with existing Dolby Digital gear in tens of millions of homes, Plus converts to DD at 640kbps through optical and coaxial outputs. While that may be a downconversion, it is higher than the previous 448kbps maximum of Dolby Digital, so if the source is 640kbps or better, the difference should be audible—at those data rates, every additional bit helps.

All of the digital surround encoding formats discussed above are lossy formats. That is, they use compression (or more precisely, data reduction) to send lots of data through narrow channels. As digital pipelines grow—most recently, with the freshly arrived Blu-ray disc format—it becomes possible to use lossless encoding to deliver more data with no sacrifice in fidelity. What you hear is a bit-for-bit reconstruction of the original master signal with *nothing* lost. That is the promise of **Dolby TrueHD** and its competitor DTS-HD Master Audio (the latter is discussed in the next section).

Dolby TrueHD, like Dolby Digital Plus (above), is a fully discrete 5.1-, 6.1- or 7.1-channel format on Blu-ray though theoretically it can support up to 14 full-range channels. It musters data rates up to 18 megabits per second—more than 40 times the data rate of Dolby Digital—for what Dolby calls "studio master" quality. In addition to sheer size, the TrueHD bitstream is also smarter. It uses metadata, or data about data, for dialogue normalization (to even out volume levels among sources) and dynamic range (to make the range of soft-to-loud more adjustable).

A key potential use of Dolby TrueHD in Blu-ray is to replace **uncompressed PCM**. PCM is the common language of many digital audio formats. The CD, for example, is a PCM format with strings of 16 zeroes and ones repeated 44,100 times per second. That makes CD a 44.kHz/16-bit format. Richer PCM audio as high as 96/24 or 192/24 is being encoded in Blu-ray releases without compression, but this is a vast waste of space. TrueHD delivers the same quality in 25-50 percent as many bits. That leaves more space on the disc for better picture quality or extra features with no sacrifice in audio performance.

Blu-ray mandates the original Dolby Digital at an improved 640kbps, but makes TrueHD (18Mbps) and Dolby Digital Plus (1.7Mbps) optional, which means the player may or may not have TrueHD/Plus decoders. In practice, most now do.

In the latest generation of Blu-ray players and surround receivers, the decoder has been added to the receiver. Some call this the **audiophile mode** because it passes the undecoded bitstream directly, instead of a converted high-res PCM or analog signal, from player to receiver through the **HDMI 1.3** interface—look for that on spec sheets. With the decoder in the player, an HDMI 1.1 or 1.2 connection will provide full performance with Dolby Digital Plus and TrueHD. If the player has high-quality digital-to-analog converters—as some early ones are reported to have—then an analog connection will also provide high performance, as long as your player or receiver has bass management to handle the crossover between satellite speakers and subwoofer. Using old-style optical or coaxial digital outputs will deliver to the receiver a downconversion to Dolby Digital at 640kbps, still higher than DD's original 384-448kbps in standard-def DVD, with your receiver's usual bass management.

Support for Dolby TrueHD and Dolby Digital Plus in Blu-ray players was until recently a moving target. You can't assume that older players handle both formats (and their DTS equivalents). The hardware lagged the software for a time, so investigate a player fully before you buy. Does it output the bitstream via HDMI? If not, does it output the new codecs as high-res PCM? Or through the analog jacks? Does it do so only after a firmware upgrade? Or not at all? The wise surround-savvy consumer will do his homework before buying an older Blu-ray player. Current-generation models are not as problematic.

Another new Dolby innovation is **Dolby Volume**. Its dual purposes are to smooth out volume levels among sources, and to tame the extremes of movie soundtracks. The reference level used to calibrate surround systems is too loud for many listeners. But when you listen at lower volumes, your ears can't hear bass and treble frequencies as well. Dolby Volume adjusts these frequencies dynamically while keeping background sounds stable, so you won't hear "pumping" effects. Dolby is licensing it for surround receivers and other products. Similar technologies include **THX Loudness Plus**, **Audyssey Dynamic Volume/EQ**, and **SRS TruVolume**.

The latest from DTS:
DTS-HD High Resolution Audio, DTS-HD Master Audio

Let's climb the DTS family tree: We've already discussed **DTS 5.1** and **DTS-ES** (6.1) above. They operate at 768kbps-1.5Mbps. Their sampling and bit rates, unmentioned above, are 48kHz and 24 bits. There is also a **DTS 96/24**, unmentioned above, used as an alternate soundtrack on some DVD-Audio high-resolution music discs that doubles the sampling rate.

You may see programming labeled **DTS Encore**, **DTS Encore ES**, or **DTS Encore 96/24**. They are identical to the three listed above. DTS uses the Encore designation to distinguish its older formats on program packages. Historical footnote: The DTS codec has an alternate name, **Coherent Acoustics**, but it's never used to label hardware or software, so ignore it. That brings us up to date on the old DTS and its nomenclature, old and new.

These older (but still relevant) standards use what the DTS people call **core data**. But there are two newer standards that add **extension data** to vastly increase performance. The core and extension data are encoded and delivered in a single bitstream. Older surround hardware decodes only the core data. Newer surround hardware can decode the extension data, accessing two new DTS standards.

One of these, roughly comparable to Dolby Digital Plus, is **DTS-HD High Resolution Audio**. It encodes up to 7.1 channels at 96/24 resolution and data rates up to 6Mbps on Blu-ray. And it is lossy, which means it uses compression and perceptual coding, just like abovementioned members of the DTS family.

But the ultimate DTS is lossless, using no perceptual coding, and omitting not a single bit of the master source. This is **DTS-HD Master Audio**. It encodes up to 5.1 channels at sampling rates of 192kHz and up to 7.1 channels at 96kHz and data rates of up to 24.5Mbps on Blu-ray.

There is a catch: The bitrates for these new members of the DTS family are variable. This allows them to kick in more bits when the soundtrack becomes more challenging—with an explosion, perhaps, or the roar of a crowd. The formats share a constant bitrate of 1.5Mbps.

How high you're allowed to climb on the DTS family tree, qualitatively, is a matter of hardware, software, and connections. Except for DTS 5.1, all the DTS standards are optional in Blu-ray. However, DTS has been fairly explicit about connectivity. For DTS-HD Master Audio you'll need the latest HDMI 1.3 (or higher) interface in both player and receiver. DTS-HD High Resolution Audio travels at up to 6Mbps via HDMI 1.1 and 1.2 (and higher). Software labeled DTS Encore (DTS 5.1 and DTS-ES) can travel by HDMI 1.1 and 1.2 or through standard old-style optical and coaxial digital jacks. And any of them operates, at full resolution, through analog jacks.

Hardware and software notes: As with the next-gen Dolby formats, DTS-HD may not be fully supported in all Blu-ray players. Some output DTS-HD MA as lower-res DTS core, even after firmware upgrades, which is very disappointing. But the latest generations of product have caught up, finally providing the full potential of DTS-HD.

DTS-HD Master Audio is now the dominant lossless surround co-dec in the Blu-ray format, with studio support far surpassing Dolby TrueHD. And it sounds fantastic.

DVD-Audio and SACD

In addition to Dolby TrueHD and DTS-HD, there are two music-only surround formats that transcend the mundane. DVD-Audio and SACD were invented to provide a higher quality medium for audio-only use. Machines handling both formats (plus DVD-Video and sometimes Blu-ray) are called **universal players**. For compatibility, your surround receiver or preamp-processor will need HDMI (1.2 or above). As an alternative, you can have the player output a high-res PCM signal to the receiver. Most receivers lack DSD decoding but the high-res PCM workaround provides about the same quality. Still another alternative is a set of 5.1-channel analog line inputs to mesh with the player's 5.1-channel analog line outputs. (There are also DVD players, receivers, and surround processors from Denon, Meridian, and Pioneer that use their own proprietary 5.1-channel digital interfaces.) For more information, see the chapter on "Picture & Sound Sources/Disc players," which devotes a section to DVD-Audio and SACD.

THX certification

THX started out as a program to correct inadequacies in film playback, ensuring that surround sound in theaters meets the standards set by filmmakers. Then it expanded into home theater with a product-certification program that makes it possible to implement Dolby Digital, DTS, and other surround formats in a more uniform manner. The home THX program, like its cinematic counterpart, ensures that home theater audiences hear what the filmmaker intended.

Dolby's and DTS's licensing requirements are designed to allow their surround formats to fit in a variety of products. THX's licensing requirements are designed to fit Dolby or DTS technology into a high-end home theater system. The aim of THX is to allow the home theater audience to hear what the engineer heard on the mixing stage, with plenty of high-impact bass and enveloping surround sound.

In recent years THX has expanded from product certification to the development and licensing of surround encoding formats. THX was a co-developer, with Dolby Labs, of the seven-speaker Surround EX format and took the lead in licensing that technology. (EX is now licensed by both THX and Dolby.)

Even as the cinematic THX program continues to cover commercial moviehouses and mixing facilities, the home THX certification program now covers everything from speakers to receivers, pre-pros, power amps, Blu-ray and DVD players, DVRs, and things like speaker cable, interconnect cables, and DVD movie releases. There are even THX-approved PCs, sound cards, and multimedia products.

THX stands for Tomlinson Holman's eXperiment, after the audio scientist who originated it while working for Lucasfilm. It's also a reference to *THX 1138*, the first film of George Lucas, whose Lucasfilm company originally controlled the development and licensing of THX.

THX comes in several forms. **THX Multimedia** is for desktop use, with the screen two feet, four inches away. **THX Select** is the kind you're most likely to see in speakers and DVD players, while **THX Select2** is for receivers. This version of THX is designed to work best in smaller rooms, with the screen 10 to 12 feet away. The original THX spec is now called **THX Ultra**, and found in surround preamp-processors as well as high-end receivers and other products. THX Ultra

107

is suitable for larger rooms, with the screen more than 12 feet away, and is different in other subtle ways. It is being phased out in favor of **THX Ultra2,** a 7.1-speaker extension of Ultra that is suitable for multi-channel music as well as movies. Products licensed to include THX Loudness Plus are certified as **THX Select2 Plus** or **THX Ultra2 Plus**. All varieties of THX require certified gear to play at high volume levels, to disperse sound in specific ways, to maintain a low level of noise and distortion, and to behave in a stable and predictable way.

Here are some of the major aspects of THX certification:

- re-equalization
- timbre matching
- rear-surround decorrelation
- bass management
- amplifier specs
- reference level
- THX Optimizer/Optimode

Re-equalization compensates for the differing tonal balances that are appropriate in home theaters versus moviehouses. It can reduce the "brightness" of many film soundtracks. In a moviehouse, the soundtrack has to have a somewhat brighter treble to enable the audience hear all the dialogue. But in a home theater, which doesn't have the hundreds of bodies and other absorptive elements of a moviehouse, that same soundtrack may seem too forward, abrasive, and oppressive. Re-EQ is so useful that it's become available even in non-THX gear in both officially licensed and unlicensed variants.

Timbre matching attempts to match the tonal qualities of the front and rear speakers to create a more uniform surround soundfield.

Decorrelation of the surround channels is a clever way of introducing a little diversity to the single mono surround channel that serves the two rear speakers in Dolby Pro-Logic surround. Why bother? Because a mono signal precisely centered between two surround speakers can collapse the channel (and thus the part of the soundfield) closest to the listener. The effect of decorrelation is to make the soundfield in the back of the room more diffuse, which is a good thing. Newer THX products have **adaptive decorrelation**, which recognizes when the circuit is not needed—in other words, when fully discrete surround channels come into play.

Bass management is (in part) a generic term for the crossover settings that a surround processor uses to route bass from the front and surround speakers to the subwoofer. THX recommends using a subwoofer and setting all other speakers to "small" during system setup. This allows you to place the sub where it sounds best in your room. When larger full-range speakers (bookshelf or bigger) are used with a sub, the sub is usually shut down. One of the great benefits of THX is that the spec is very consistent about bass management, in terms of both the 80Hz crossover point (the frequency at which the sub takes over from the main speakers) and the slope (the way one driver's response declines as another driver takes over). In some products, bass-management options go beyond the basic crossover to include additional subtleties. For instance, you might have the option of mixing bass information from the front left and right channels with the sub channel.

Amplifier specs are tougher in THX gear than in non-THX gear. A THX-certified receiver or power amp must be able to drive a speaker with an impedance of 3.2 Ohms (that's a very demanding load) and produce a volume level of 105dB (that's very loud). That will handle just about any movie-soundtrack or musical peak known to humankind. THX Select2 equipment must achieve this feat in a room of 2000 cubic feet (height times width times depth). THX Ultra2 equipment must do it in a room of 3000 cubic feet. The room-size numbers are a guideline, not a dictate, but it certainly pays to know how big your room is when shopping for THX gear.

The THX **reference level**—synonymous with the Dolby reference level—is intended to replicate the level at which the movie was mixed. Using it will keep dialogue at a moderate level, while effects and music may have more dynamic punch. Both Dolby and THX specify a reference level of 85dB (decibels). At the **controller** (THX-speak for pre-amp-processor), the volume meter registers the number of decibels above or below reference level (plus or minus one, etc.).

THX Optimizer, or **Optimode** is for DVD movie releases. It places test patterns on the disc to enable the home theater buff to tweak the video display to the same settings used by the mastering engineer. This is a brilliant and useful idea.

THX Ultra2 certification

Here are some of the major aspects of THX Ultra2 certification in addition to the basic THX requirements listed above:

- EX auto detection and switching
- high-bandwidth video switching
- adaptive side-surround decorrelation
- boundary gain compensation

EX auto detection and switching detects a flag in EX-compatible DVD-Video releases and switches on the EX back surrounds. When the flag is missing, the user may switch on the EX circuit manually if desired.

High-bandwidth video switching requires that a THX-certified device be able to pass high-definition and/or progressive-scan video signals without degrading them. THX does not specify the necessary bandwidth but does test licensed products to ensure that no degradation is visible.

Adaptive side-surround decorrelation switches itself on when the surround processor receives mono surround signals (for example, in Dolby Surround-encoded software). It affects only the side-surrounds, not the back-surrounds. When the processor receives stereo surround signals (for example, in Dolby Digital 5.1-encoded software), they are sent to both the side-surround and back-surround channels, and decorrelation is switched off. This assumes that the movie has stereo surrounds and that EX is switched on.

Boundary gain compensation is a switchable circuit that tames excessive bass. Applied before the processor's bass management circuit, it is intended for listeners sitting near the rear wall where acoustic conditions tend to result in exaggerated bass response. Many rooms are not large enough to allow placement of a sofa in the middle of the room. If your sofa is placed near the rear wall, boundary compensation will provide more accurate bass for listeners sitting there.

The latest from THX

THX has always been oriented primarily to audio. But in 2006, that began to change with the company's first certifications of video products, including Runco projectors and TiVo DVRs.

As with any kind of THX certification, manufacturers submit products in the prototype stage for testing. In video products, THX testers are on the lookout for motion artifacts, ghosting, dropped frames, and noise. Their front-of-screen testing covers brightness, contrast, color gamut (range), color gamma (balance), uniformity, and resolution. Video signal processing tests include scaling, deinterlacing, and conversion, especially as it affects motion.

Whether THX certification will catch on in the video industry (and videophile community) remains to be seen. However, in a marketplace where the performance of products varies so wildly, there certainly is a need for something like it.

Another strategic shift for THX, starting in 2007, is a deeper involvement in product design. THX already maintains labs to certify products. Increasingly, manufacturers who are beginning to outsource their design are turning to THX for help with mechanical, electronic, and firmware design.

THX Loudness Plus, a new innovation circa 2007, recognizes that the reference level is too loud for many listeners. It compensates for tonal and spatial shifts that occur when you listen at lower volumes. It is available in THX Ultra2 Plus and Select2 Plus certified products. Other licensed versions of similar technology include **Dolby Volume** and **Audyssey Dynamic Volume/EQ**.

A project code-named **Blackbird** would take THX to a new level by using metadata—or data about data, similar to the way your iPod keeps track of artist and song names—to adjust aspect ratio, volume reference level, picture settings, and surround modes. Blackbird will make our lives easier when it begins arriving in certified products.

In 2010 THX announced an initiative aimed at the nascent 3DTV category. In conjunction with BluFocus, THX is preparing three new categories of certification. **THX-BluFocus AV Certification** would ensure that sound and image measure up to the master source. **THX-BluFocus Creative Certification** would ensure that 3D elements not deviate from the director's intent or cause viewer fatigue, Finally, **THX-BluFocus Interoperability** would certify that discs play on both 3D and 2D players.

To learn more about the THX certification program, visit thx.com.

Audyssey's auto-setup and listening modes

In addition to Dolby, DTS, and THX, there is another emerging player in surround technology. Audyssey was founded by R. Tomlinson Holman and Chris Kyriakakis of USC—Holman had previously co-founded THX. Audyssey licenses auto setup, room correction, and low-volume listening modes as well as other listening modes that add width and height enhancements to the basic 5.1-channel configuration of surround.

Audyssey got its start by licensing combined auto-setup and room-correction technologies to makers of surround receivers. These make it easy to set up a receiver by measuring the speakers and room, then making adjustments for speaker size, speaker distance, and room acoustics. Versions start with **Audyssey 2EQ**, which takes measurements from up to three positions in the listening room, and provides "basic resolution filters" for satellite speakers but not for the sub—all other versions filter both sats and sub. **Audyssey MultEQ** measures from up to six positions (or 32 if you're a custom installer) with "mid-level resolution filters." **MultEQ XT** measures from up to eight positions (32 for CI), with "high-resolution filters." The latest version, **Audyssey MultEQ XT32**, makes the best of the limited audio processing powers in receivers, using higher sampling rates where they will do the most good, and providing 32 times as much filter resolution as its predecessors. Audyssey is the class act of its field—if you want auto setup and room EQ, Audyssey is an excellent choice.

When a system is running at reference level, the dynamic extremes of movie soundtracks can be fatiguing. You turn the volume down to avoid the loudest effects, then turn it up again to catch dialogue, and this quickly becomes tiring. Audyssey is among the companies licensing **low-volume listening modes**. **Audyssey Dynamic Volume** smooths out volume variations, while **Audyssey Dynamic EQ** adjusts frequency response and surround envelopment, so that soundtrack elements are not only audible but natural-sounding at low volumes. Turning on Dynamic Volume always activates Dynamic EQ; but when Dynamic Volume is off, Dynamic EQ can be turned on or off. So you can use Dynamic EQ without being bound by Dynamic Volume's level-setting decisions. Other companies licensing low-volume listening modes include Dolby (with **Dolby Volume**) and THX (with **THX Loudness Plus**).

More recently Audyssey has moved into surround post-processing modes with **Audyssey DSX**. This is the first listening mode to add width channels, and the second (after **Dolby Pro Logic IIz**) to add height channels. What is the benefit of adding width channels? According to Audyssey, the ear can hear in more directions than current technology allows. "Experiments have shown that human localization is better in front than to the sides or behind. This means that for front-weighted content such as movies and most music, good engineering dictates that we employ more channels in the front hemisphere than the back. Imaging is also better horizontally than vertically and so good en-

112

gineering also dictates that channels must first be added in the same plane as our ears before going to higher elevations." What Audyssey is saying is that, if you've got seven amp channels to work with, adding width will provide the greatest benefit, adding height is second best, and adding back-surrounds is third best. In my opinion, both width and back-surrounds are dispensable, while height can be fun for some movies. None of them invalidates an existing 5.1 system—the incremental benefits are minimal. See the "Installation guide" for placement tips.

Minor proprietary surround modes

The following modes are found only in a handful of products. These are not mainstream surround formats that involve encoded software and decoding hardware, like the Dolby and DTS formats, though they may be fun for a few listeners.

Logic 7 derives 7.1 channels of sound from 5.1-channel or stereo sources. It is found in receivers by Harman Kardon and high-end surround processors by Lexicon.

Circle Surround, from SRS Labs, adapts stereo sources to surround. The 5.1-channel version competes with Dolby Pro Logic II while the 6.1-channel version (**CS II**) competes with DPLIIx and DTS Neo:6. Found in Marantz receivers, Circle Surround promises greater "dialogue clarity" and "cinema-like bass enrichment." It has a brighter and more forward sound than the other modes, making it suitable for speakers that have a relaxed, laid-back sound.

Cinema DSP is Yamaha's name for a mode that adds two front effects channels to a 6.1-channel array. That means that the total of *front* channels rises from three to five. Yamaha receivers with this mode have a grand total of eight channels plus sub.

DSP modes

To complement the mainstream surround formats, most surround receivers and processors offer **DSP** (digital signal processing) or **EQ** (equalization) modes that include various surround ambiences (sometimes plus or minus a few channels). Some receivers have dozens of them. These are intended primarily for processing stereo music sources. However, with the advent of Dolby Pro Logic II, DSP modes have become more tangential.

Overdone DSP surround effects sound "wet"—dripping with re-

113

verb. They're entertaining for a few minutes but their crudity quickly becomes tiring. More subtle ones send only a whisper of ambience to the rear-surround speakers and are more listenable over the long haul.

DSP is also used to process mainstream surround formats. However, the phrase "DSP modes" usually refers to non-licensed additions.

Are five channels too many for you?

A system can have more than 5.1 channels or fewer than 5.1 channels. More than just exercises in reductionism, the latter are valid ways to build a practical surround system—by adapting technology to your needs, not vice versa. (Note the two forms of nomenclature: mine first, Dolby Labs second. With Dolby's method, the sub is implied.)

Surround 5.0 (or 3/2) keeps the three front and two surround speakers, but eliminates the subwoofer, routing bass to the other speakers. Set all speakers to "large," subwoofer "off." If two front speakers (or all five) are full-range tower or bookshelf models, why bother adding a sub, especially if you find action-movie dynamism more an irritation than an asset? The classic stereo pair can provide adequate to excellent bass. Using big speakers in lieu of a subwoofer is a far better option than buying a mediocre sub. But you will need either speakers with a high sensitivity/efficiency rating or a muscle amp to drive more demanding speakers.

Surround 4.0 (or 2/2), with two channels in front and two in the rear, was the original configuration of **Dolby Surround**, as it was then called. This was before Pro-Logic added the front-center channel, and way before Dolby Digital. You don't have to dig up vintage Dolby gear to produce 4.0 surround. Just tell your surround processor that there's no front center speaker; or, if you're using an old Pro-Logic processor, switch the front-center channel to **phantom** mode. In either case, this effectively shuts down the center speaker and redistributes its signal to the front left and right speakers. This might temporarily be a better alternative than leaving a center speaker atop a TV where a busy toddler might tug on its cable. Or you may not have an adequately matched center speaker. Or there may not be a position where a center speaker blends in acoustically—spaces can be unpredictable.

Surround 3.0 (or 3/0) uses the **Dolby 3 Stereo** mode found in some surround processors, though you can achieve the same effect by simply shutting down the surround pair but keeping the front left, center, and right speakers. Ditching the surrounds may be a necessity in some room layouts, especially those where speaker stands, invasive wall mounting, or long cable runs may not be practical (even a tiny screening room can easily require dozens of feet of cable to reach each rear speaker). Surround 3.0 provides all the benefits of stereo while eliminating the traditional "hole in the middle" weakness of two channels. If you have Dolby processing gear and three matching speakers, you can have a strongly continuous front soundstage. And to take advantage of it, there's lots of stereo on the air as well as on tape and disc.

Surround 2.0 (or 2/0) is stereo. There are still a few audiophiles (old and grey and very stubborn) who insist 2 channels work better than 5.1. Surround 2.0 goes beyond the embrace of plain-vanilla stereo to include enhanced stereo formats that simulate surround effects from two speakers. Examples include Hughes' **SRS** (Sound Retrieval System), which has appeared in everything from TVs to laptops; **Spatializer,** in several brands of DVD players; and **VMAx,** in Harman Kardon receivers. Dolby Labs has added its own **Dolby Virtual Speaker**, which works surprisingly well, while DTS offers **DTS Surround Sensation** for PCs.

Picture & Sound Sources

*A*system is only as good as what it's fed. The mistaken belief that all signal sources perform the same is far more entrenched in the home theater sphere than in the high-end audio world. We home theater buffs could well afford to take an attitude lesson from our audiophile friends who take pains to feed their systems the highest-quality signals available from turntables or digital sources. Choosing the correct picture and sound sources also has ramifications in ease of use, as anyone who has ever struggled with a badly designed disc player will quickly comprehend.

Right now the highest-quality sources of digital television signals are Blu-ray, which supports 1080p, and the airwaves, where HDTV in the 1080i and 720p formats is being broadcast. Most cable systems are now required to offer digital versions of local channels. DirecTV and the Dish Network offer packages of HDTV channels. The DirecTV version of HDTV is slightly truncated but still high-quality. D-VHS VCRs and an increasing number of HD-enabled, hard-drive-based digital video re-corders are capable of capturing the full image resolution of HDTV (though regular VHS VCRs and some DVRs are not). In the standard-def domain, progressive-scan DVD is the best choice, followed by conventional interlaced DVD and (if you're in a nostalgic mood) laserdisc.

For users of analog TVs, DVD is the best choice, followed by satellite (on a good day), laserdisc, broadcast (with good reception), cable, and VHS in last place. Though people still using analog TVs probably don't care.

Disc players

There are plenty of ways to get movies and music into a home theater system, but the primary hard-copy video formats in home theater are high-def **Blu-ray** and standard-def **DVD** (which stands for **Digital Video Disc** or **Digital Versatile Disc**, depending on whom you ask). DVD has so far been the fastest-growing format in the history of consumer electronics, exceeding both the VCR and the CD. Its high-def replacement, **Blu-ray**, is growing far more slowly, and will not be knocking the original DVD format off its throne for awhile. Blu-ray is available in standalone players, PlayStation3 game consoles, compact in-a-box systems, PC drives, and a few HDTVs—and no doubt it will get more versatile over time. It's also penetrating hitherto DVD-only territory such as **jukeboxes**, five-disc **carousels**, and portables.

Blu-ray: the future of high-def on disc

The long-awaited high-definition videodisc has arrived! Unfortunately, it arrived in duplicate, as the consumer electronics industry shot itself in the foot yet again with yet another childish squabble over licensing revenue. Remember Beta vs. VHS? The latest skirmish has been Blu-ray vs. HD DVD, with Blu-ray emerging as the victor.

Let's briefly pay tribute to HD DVD before we dismiss it from our minds. Toshiba and Warner—two of the most significant developers of the original DVD—invented a format that harnessed advanced video compression to a new type of blue-light laser (existing DVD uses a red-light laser). This **HD DVD** format stored 15 gigabytes on a one-layer read-only HD DVD, 30GB on a dual-layer disc, and 51GB on a triple-layer disc. In its rewritable form, it stored 30GB on one layer, 60GB on two, or 102GB on three. Among other advantages, HD-DVD was approved by the DVD Forum, the standard-setting consortium that keeps the DVD flame alive. And it used a disc manufacturing process similar to (and therefore as cost-effective as) the original DVD-Video format. Blu-ray discs require a new and more costly manufacturing process.

117

But other companies, notably longtime format warrior Sony, didn't want to pay licensing fees to Toshiba and Warner. The Blu-ray Founders Group (since renamed the Blu-ray Disc Association) counterattacked with a competing **Blu-ray** format. It also uses a blue-light laser to store up to 25GB on a single-layer disc, or up to 50GB on a dual-layer disc—significantly more than HD DVD, as Blu-ray partisans never tire of pointing out. Future versions may increase capacity by multiplying the number of layers. Initially Blu-ray had an Achilles heel. Because it read the disc at 0.1mm—a much shallower depth than HD DVD (and existing DVD), at 0.5mm—the disc was fragile, and had to be enclosed in a cartridge, evoking sad memories of the deceased CED videodisc from the 1980s. However, TDK saved the day by developing a hard disc coating that eliminated the need for the cartridge.

So, for those video historians who would like to split hairs, which has the higher picture quality—Blu-ray or HD DVD? Both use video compression: either MPEG-2, MPEG-4 AVC, or VC-1 (see "DTV by the numbers"). The answer lies in the encoding—that is, how many bits the mastering engineer allocates to the image and how he chooses to manipulate them. Theoretically, Blu-ray has a slightly larger bit bucket, but both formats have the potential to look amazing. Both support trendy 1080p and 24fps video output. But only one survives, and that's Blu-ray.

On the audio side, Blu-ray mandates Dolby Digital at improved data rates up to 640kbps, plus DTS and uncompressed PCM. Some discs include stereo soundtracks as well as surround. Optional audio codecs include the new lossless codecs, Dolby TrueHD and DTS-HD Master Audio; and the improved lossy ones, Dolby Digital Plus and DTS-HD High Resolution Audio. DTS-HD MA has proven to be the choice of most studios, firmly establishing lossless surround in most Blu-ray disc releases. Players may also downmix surround soundtracks for output via stereo analog jacks.

With the Blu-ray Founders Group including Hitachi, LG (Zenith), Panasonic, Pioneer, Philips, Samsung, Sharp, Sony, TDK, Thomson (former parent of the RCA brand), and even Toshiba as of 2009, the Blu-ray format has all the support it needs, especially since most of these companies are actively engaged in manufacturing Blu-ray products. Sony and Panasonic have been selling high-priced Blu-ray recorders in Japan, recently adding higher-capacity versions. The first Blu-ray player to arrive in the United States, in mid-2006, was a Samsung product that was marred by faulty video processing circuitry that rendered it sub-high-definition though that problem was corrected. Blu-ray players are now available from many manufacturers. All major studios—including some formerly pledged to HD DVD—support Blu-ray with software releases. That includes not only movies but videogames from Electronic Arts and Vivendi Universal.

Blu-ray's interactivity standard is **BD Java**. The original and now outdated version was Profile 1.0. Profile 1.1 supports **Bonus View** features such as expanded local storage, picture-in-picture, and internet access. It's also called Final Standard Profile, which is confusing, because it's far from the final one! Profile 2.0, a.k.a. **BD-Live**, players support internet-enabled features such as chat and messaging. But the Blu-ray camp didn't designate it until October 31, 2007, and some manufacturers rush-released players to beat the deadline. So some Blu-ray players out there won't handle BD-Live features on existing and future disc releases because the hardware lacks the requisite on-board memory and ethernet jacks. Buy BD-Live, Profile 2.0. (This is one area where HD DVD did a better job from the start, by choosing Microsoft's **HDi** format from the outset. Ethernet connectivity was mandated in all players, making it easy to upgrade firmware. Not that any of that matters now.)

Be warned that these next-generation players have some surprising limitations. Amazingly, some older Blu-ray and BD/HD DVD combi players don't play CDs. If you want one player to rule them all, check spec sheets before you buy. You'll also find that older Blu-ray players take a long time to boot up and an additional long time to load disc

Oppo

The Oppo BD-83 ($599) is a universal disc player for the age of Blu-ray. Besides BD, it also plays SACD, DVD-Audio, DVD-Video, CD, and other formats. It is getting spectacular reviews for its performance, ease of use, build quality, and speed.

menus. And their fast-scan modes can be slow to respond or balky. Fortunately, newer models are much more speedy in every respect, though some are better than others. Check load times when you shop.

For digital rights management, Blu-ray and HD DVD both use a scheme called **AACS** (**Advanced Access Content System**). Among many other things, this permits encryption that could cause an "attacked" player to shut down. AACS has already been hacked several times, with encryption keys for several HD DVD releases posted on the net. Blu-ray adds two additional forms of protection. **BD+** can examine and verify player keys, or alter the player's output. It also has been hacked. The **BD-ROM Mark** is a cryptographic mark that is stored separately from other content and is needed to unlock playback. So far it has not been hacked.

AACS also includes a flag called the **image constraint token** that would down-res high-def signals to standard-def when delivered through the less secure component video outputs. However, following videophile howling, the studios have agreed not to implement the ICT in software until December 31, 2013, after which all players manufactured will have to down-res via component. If you're upgrading your player, buy one made before that date.

One of the coolest new Blu-ray and DVD features is **network connectivity**. Some players with ethernet connections can access media from a PC through a router. **DLNA-certified players** can access music, video, and photos via the Windows Media Player.

Even in the wake of HD DVD's demise, Blu-ray still has some faint competition. A third, and minor, entry in the high-def disc war was **HD VMD**, which stood for High Definition Versatile Multilayer Disc. Developed by New Media Enterprises, the disc was read by a red (not blue) laser. A few Bollywood titles were released but the format is on ice now.

In addition to these formats, other high-definition disc formats materialized in the Far East, including China's **CBHD**. It uses a blue laser to read a 30GB disc and uses some technology inherited from HD DVD, thus allowing the developers to avoid playing licensing fees to the Blu-ray people. The audio/video codec is the homegrown Chinese **AVS**. Warner announced software support in 2009. CBHD is now outselling Blu-ray in China.

A few HD movie titles have been released in the **UMD** (**Universal Media Disc**) format developed by Sony for the PlayStation Portable gaming console. The format has not caught on.

Many manufacturers, starting with Samsung, have introduced DVD players that upconvert their standard-definition output to high-definition formats. These are not true high-definition players but they may provide a smoother picture with some high-definition displays. Toshiba's **XDE** (**eXtended Definition Enhancement**) players are so sophisticated that you may be able to convince the neighbors they're watching Blu-ray, unless they are knowledgable viewers.

For users of Windows PCs, Microsoft has supported HDTV in 1080i and 720p, with 5.1-channel surround, in Windows Media Player starting with **Windows Media 9**. The first DVD release to use WM9 was the second disc of the two-disc set of *T2: Extreme DVD* in June 2003. Windows Vista and Windows 7 support Blu-ray.

Another largely PC-based high-def video compression format is **DivX**, from DivX Networks. While it is available in players from at least a half-dozen major brands, there is no prerecorded software in the format, so you'd be limited to PC-generated discs, presumably containing downloaded material—not all of it legal.

All of these would-be successors to the wildly popular standard-definition DVD format may eventually be overtaken by new formats using **holographic laser** technology. It uses multiple lasers to read and write data three-dimensionally, as opposed to addressing a single pit. And it promises to store far more data than any technology using a single red or blue laser. Its most likely use will industrial, not consumer.

The DVD-Video format

The **DVD-Video** format (most just call it DVD) is a variation of the physical read-only disc format known as **DVD-ROM**. As originally designed, DVD-Video delivers **MPEG-2** compressed video of varying quality, combined with up to 8 streams of sound usually encoded in **Dolby Digital AC-3**. (MPEG is a standards-setting outfit, the Moving Picture Experts Group; AC-3 is the original name of Dolby Digital.) However, with the growth of the DVD-Video format, the selection of possible soundtracks has become more complex.

If the movie is a contemporary one, six channels (or 5.1, to use the usual nomenclature) will form a surround soundtrack. The other two may provide mono, stereo, or two channels encoded with Dolby Surround. Some discs include 5.1-channel Dolby Digital *and* DTS soundtracks (sometimes accompanied by a reduction in video quality as the **bit bucket**—the supply of digital data—runs low). These dual-5.1 discs may omit the two-channel alternate soundtrack. If no two-channel sound-track is available for use with a stereo system, the DVD-Video player can mix a 5.1-channel soundtrack down to stereo. This is called the **Dolby Digital mixdown**. (None of this has anything to do with **DVD-Audio**, a separate 5.1-channel audio-only format that provides for superior audio quality.)

MPEG-2 is a form of **lossy** video compression used to fit movies within the storage capacity of a disc. When films are mastered for DVD release, the MPEG-2 encoder pares down the digital video information to eliminate redundant data such as backgrounds that don't move. When the encoding is done well, its effects are invisible. When it's done poorly, a variety of distracting digital **artifacts** result, especially in scenes containing rapid movement or other complex video information. Common artifacts include jagged diagonal lines (**jaggies**) and a picture dissolving into blocks. Video displays can introduce artifacts of their own.

DVD-Video is not an **HDTV** (high-definition television) format. To get HDTV look to free broadcasts, satellite, cable, HD DVRs fed by those sources, Blu-ray, and D-VHS. Even within its standard-def limits, DVD's MPEG-2 video compression varies according to the skill of the mastering engineer, who allocates the disc's storage capacity. Scenes with rapid movement all over the screen take up more disc space than those with unmoving backgrounds. Another factor is the quality of the authoring software. If you thought one movie looked great while another looked awful, it's probably not your equipment that's at fault.

Quirks notwithstanding, DVD does offer a clean, colorful picture that has thrilled home theater audiences for more than a decade now. Not only is the picture sharper than that of VHS tape—it is also more stable, something that becomes immediately obvious when you're trying to read the fine print in movie credits (especially those fast-moving song titles). That picture comes in either of two officially sanctioned **DTV** (digital television) formats known as **EDTV** (enhanced-definition television) and **SDTV** (standard-definition television). Either way, DVD produces a picture with vertical resolution of 480 pixels and horizontal resolution of 720 pixels, roughly the limit of most conventional analog TVs, though many DTVs can do better. The incoming Blu-ray format supports resolutions up to 1080 by 1920.

On most TVs, including analog sets, virtually any DVD-Video player will produce a better picture than a VHS VCR, and an adequate DVD player doesn't cost much (though a really good one costs more). However, those using digital TVs might want to pay more for a progressive-scanning model (see separate section)—or better yet, get Blu-ray.

More DVD-Video features

The DVD-Video format is a source of wonder to anyone who grew up with the boring old VCR. For one thing, using a DVD-Video player is more reminiscent of CD than VCR. Instant random access lets you flit from scene to scene—assuming **chapter stops** have been inserted at the most helpful moments by the mastering engineer. Some releases make a big deal of chapter stops, with illustrated menus. Special effects are generally smoother than with a VCR.

Single-sided DVD-Video discs can hold up to 5.2GB (gigabytes). That does not mean all DVDs hold that amount; capacity varies with the number of layers and sides. The format supports both **dual layer** discs (the laser briefly delays the program to refocus from one to the other) and **dual-sided** discs (the user must flip the disc). A dual-sided disc sometimes may have a widescreen version of a movie on one side and a nonwidescreen version on the other.

There's enough storage space to accommodate more than one version of a movie, soundtracks in different surround formats, foreign language soundtracks, captioning in multiple languages, up to nine camera angles, seamless branching of varying storylines, "making of" documentaries, and still photos. You might even find the hidden features that some buffs call **Easter eggs** (mostly trivia-related video clips). All players support these features—but disc releases vary. Use of the multi-angle feature, for instance, is still disappointingly rare.

DVD-Video and Blu-ray both include **regional coding** which permits movie studios to designate different release versions for different parts of the world (North America, Europe, Japan, etc.). This may be convenient for marketing purposes, but it frustrates collectors, who seek out doctored players that defeat the regional coding feature.

Copyright protection schemes are also encoded into movie releases. The main one for the DVD format is **CSS (Content Scrambling System)** which ensures that DVDs play back only on licensed DVD players. Also common is the **Macrovision APS** (Analog Protection System), which is designed to foil VCR-recording circuits, but also causes flashing and other unpleasant problems with certain video displays.

The **parental lock** feature is intended to prevent children from viewing unsavory material. Parents must set up a passcode. Adult holders of the passcode can view anything they want—kids can't. Of course, passcode holders must keep turning the feature on and off if they don't want to live by the same standards they impose on their kids.

A more sophisticated parental control feature is program filtering. One such technology is from the Salt Lake City-based firm **ClearPlay**. A ClearPlay-compatible DVD player can be set to use any of 14 levels of filtering to screen out violence, sex, or other material deemed offensive by some. The filters are compiled for every major new DVD release by the company's staff of "movie professionals" and can be updated by a CD sent by U.S. Mail or by downloading them from a website and burning them onto a CD-R. Other products use **TVGuardian** filtering.

Some DVD-Video features don't appear on all players and may serve as points of differentiation to the alert shopper. **Digital zoom** lets you zero in on the action (and is an excellent way to learn what digital video distortion looks like). **Picture modes** allow some players to tailor their output for different kinds of video display or different kinds of programming.

One of DVD-Video's more obscure areas of optional compatibility is the **VCD** (**Video CD**) format, which uses MPEG-1 video encoding with a fairly unsubtle form of video compression. Except as a file-sharing medium, VCD never really took off in the west. In the far east, however, it is wildly popular. On the streets of Chinese cities, a VCD player costs as little as $20 and discs go for $1. Why so cheap? Because a VCD player is very similar to a CD player, because China is virtually a patent- and royalty-free zone, and because there's an apparent need for what amounts to a poor man's DVD format. In fact, in 1998 the Chinese government developed various improved versions of VCD. Western home theater buffs will find that VCD's picture quality is not sufficient for big-screen movie viewing, but if you dare to explore the VCD format, you might find some interesting music videos and karaoke titles. Visit your local Asian enclave, where stores sell VCDs.

You really know DVD has arrived when it's spawned its first dead-on-arrival sub-format. **Divx**, which stood for Digital Video Express, was invented by Hollywood law firms, championed by Circuit City, and resented by consumers who polled more than 95 percent against the format's very existence. As long as Hollywood remains tantalized by the idea of turning home theater technology into an electronic box office, Divx will return, under many names, and one of them may beckon to consumers someday. Given the atmosphere of mutual distrust, that day may not come soon. Coincidentally there is also a compressed-video format called DivX which is an entirely separate matter.

Another attempt at a self-destructing DVD came from the Walt Disney Company. Disney's **EZ-D DVD** was made with a special resin that rendered itself unplayable 48 hours after removal from its airtight package. It was not a conspicuous success, but that didn't prevent **Flexplay** from trying again with perishable discs sold through Staples.

Another new feature that will probably be more palatable to consumers is **networking**, usually through an ethernet wired connection. This allows the DVD player to access digital video, music, or photos on a network-connected PC (though some file types may not work).

PSSST! Rumors abound concerning an underground DVD player feature quietly implemented on some players—the ability to skip ads and copyright warnings. Hit **stop, stop, play** at the beginning of a disc and see what happens. Whether it will work depends on how the disc was authored.

Audio features of DVD-Video

Before we discuss the DVD-Audio format, let's go through a few more of the optional audio formats that can enter your home through DVD-Video players.

- Dolby Digital
- DTS
- virtual surround
- CD, CD-R, CD-RW
- HDCD
- MP3/WMA

If for any reason the jaw-dropping power and efficiency of Dolby Digital does not meet your standard of excellence, another format bids to take its place—literally. **DTS** soundtracks sometimes crowd Dolby Digital 5.1 right off the disc, though Dolby Digital 2.0 (and Dolby Surround) sometimes remains for the sake of compatibility with Dolby-standard surround processors. Some complain that a fat DTS soundtrack may cause fewer bits to be allocated to video, reducing picture quality. Does DTS sound better than Dolby? Let the debate rage on; I'll have a Bombay Sapphire Gin martini, straight up, olive, please. DTS is standard equipment in audio/video receivers and DVD players. However, only a minority of DVD movie releases are encoded in DTS.

A few players add **virtual surround** formats, like Spatializer, that simulate a surround soundfield from two speakers. This faux-surround approach is no substitute for a proper 5.1-channel system but can make built-in stereo TV speakers sound a little more lively. **Virtual Dolby** surround formats are those that process signals using Dolby encoding formats but still deliver sound through two speakers.

You may not have many DTS-encoded DVDs, but if you love music, you probably have whole shelves devoted to CDs. While the DVD-Video format specs do not officially include **CD compatibility**, manufacturers provide it as a courtesy to the consumer. In a DVD/CD changer, CD capability is even more fun, allowing you to mix DVDs with CDs—perhaps a little intermission music?

Unfortunately, some DVD-Video players may not respond to commands as speedily as your old CD player did, because their laser pickups are optimized for DVD. Consumers ask a lot of our DVD players. We want them to read every kind of disc ever invented—DVD-Video, CD, DVD-Audio, SACD, etc.—and while that desire is understandable, it isn't always realistic. Manufacturers have responded by endowing some step-up models with **dual laser** pickups with an extra laser diode to read CDs. This may provide either an extra measure of disc-reading precision or speedier response to commands.

So much for storebought CDs. What about homemade **CD-Rs** and rewritable **CD-RWs**? The first few generations of DVD players handled them inconsistently, if at all. Sometimes one brand of CD-R blank would work, while another got spat out with a "no disc" error message. However, manufacturers have become more sensitive to the problem, and as a result a greater proportion of DVD players have become compatible with CD-R and CD-RW discs. For extra assurance, either look for **MultiPlay** certification (by the Optical Storage Technology Association, osta.org). Or just take a few favorite brands of recorded CD-R to the store and try 'em out to be sure.

Look carefully at your CD collection and you just might find dozens of **HDCD** releases. If that's the case, seek out an HDCD compatible DVD player or receiver. HDCD stands for **High Definition Compatible Digital**. Invented by Pacific Microsonics and now owned by Microsoft, it is a stealth format that shoehorns 20 bits worth of digital information onto a standard 16-bit CD. An HDCD sounds its best on an HDCD-compatible player but will play on any CD or DVD player. The improvement in sonics is real, if subtle, and has come with a refreshing absence of pain or hype—HDCD has not caused a format war, invalidated an older format, or raised the cost of CDs. And it's now em-

126

bedded in more than 5000 CD releases. HDCD capability is a feature in some DVD and Blu-ray players and surround receivers.

If you've spent a lot of time sharing music files over the internet, you may have quite a collection of **MP3s** by now. Want to hear them in surround sound, or hear what your subwoofer can do for that bassline? It can be very convenient to burn MP3s onto CD-R or CD-RW and play them on your DVD player. MP3 compatibility has become a standard feature. But watch out for file-name restrictions. Some players truncate file names to as few as eight characters. Also, some players may not support some MP3 data rates or other codecs, and those files won't play.

Microsoft has convinced the companies who make chips for 90 percent of DVD models to include **Windows Media Audio** compatibility in DVD players.

DVD-Audio and SACD

These two formats were invented to provide a higher-quality medium for audio-only music listening than the compact disc. Yes, we were promised CDs would provide "perfect sound forever"—and for the majority of listeners, they sound fine—but many audiophiles have been reluctant to give up their LPs, saying that CDs sound harsh, vague, grainy, or two-dimensional.

In addition to higher fidelity in general, DVD-Audio and SACD offer an opportunity to expand the two-channel palette of music recordings to as many as 5.1 channels. Both formats also support stereo recordings.

DVD-Audio and SACD are not the first formats to provide music in surround. In addition to a handful of LPs from the quad era, a handful of CDs have been released with Dolby Surround or DTS. However, only a few music producers have taken advantage of the opportunity to mix in those formats. To provide flexibility, the format specs for both DVD-Audio and DVD-Video allow the use of 5.1-channel Dolby Digital—but that audio-for-video format uses perceptual coding to eliminate supposedly inaudible audio data. DVD-Audio and SACD, in contrast, are music-delivery formats that do not eliminate any part of the original audio signal. While both (especially DVD-Audio) do support visuals, neither is intended for movie viewing.

DVD-Audio

DVD-Audio builds on the success of the DVD-Video format. DVD-Audio programming can be played only on a DVD-Audio player (not all DVD-Video players qualify). The DVD-Audio format delivers anywhere from 2 to 5.1 channels of music encoded at far higher data rates than the current CD format. Whereas the CD delivers a string of 16 zeroes and ones 44,100 times per second, DVD-Audio delivers a string of 24 zeroes and ones 96,000 to 192,000 times per second—or three to six times more bits. Depending on the requirements of programming and disc capacity, DVD-Audio offers the option of using **Meridian Lossless Packing** (**MLP**) to fit more data onto the disc. Unlike the perceptual coding scheme used in Dolby Digital, MLP reconstructs the full digital audio signal, bit by bit, eliminating nothing.

The current DVD-Audio format was preceded by a stereo format advertised as a capability in many DVD-Video models. It was known under several names including **24/96** and **PCM**. (The latter is a generic name for a type of digital encoding also used by the CD format.) This earlier format's 24-bit encoding and 96kHz sampling frequency beat the CD's 16 bits and 44.1kHz hands down, offering subtle but audible benefits. However, the full-blown version of DVD-Audio does the same thing far more efficiently thanks to MLP, and can do it in 5.1-channel surround sound, not just stereo.

DVD-Audio releases often include both uncompressed DVD-Audio and one additional compressed soundtrack that may be either Dolby Digital or DTS. Discs that include an alternate soundtrack (as most do) allow an extra measure of flexibility. If you don't own a DVD-Audio player, you can still play the alternate soundtrack on your DVD-Video or Blu-ray player. Dolby Digital and DTS are both compressed formats, and therefore won't sound as good as DVD-Audio, but having an alternate soundtrack option allows you to stock up on DVD-Audio discs while you save your pennies for a DVD-Audio player.

One drawback of first-generation DVD-Audio gear was a lack of bass-management features—in other words, the ability to designate speakers as large or small and route bass to a subwoofer. In a system with full-range speakers, a powerful rhythm section might deliver an overwhelming quantity of bass. However, newer DVD-Audio gear does include bass management. Also—addressing another early criticism of the format—newer DVD-Audio players default to the DVD-Audio soundtrack, even when used without a video display.

SACD

DVD-Audio has suffered from a format war with the **SACD (Super Audio Compact Disc)** format from Sony and Philips. SACD produces far more detail and subtlety than the CD thanks to its use of Sony's **DSD (Direct Stream Digital)** technology which records four times as much information as a regular CD. DSD originated as a reference-quality format for professional recording, mastering, and archiving. Frequency response extends to 100kHz (five times the CD's 20kHz) and the sampling rate is an astronomical 2.8 million bits per second. (A direct comparison with DVD-Audio is rather tricky due to DVD-Audio's multiple sampling rates.)

SACD has some so-called **multi-channel management** features that are reminiscent of mainstream surround formats in general. These include **channel-level adjustment** (from the listening position), **channel-balance adjustment** (front-to-back and side-to-side), and **bass management** (which can redirect bass from main speakers to sub). The format supports graphics and video but does not require a video display to be running when SACDs are playing.

SACDs can have dual layers to double running time. But the coolest thing about SACD is that it can (and sometimes does) come on a **hybrid disc** with a regular CD layer and a souped-up SACD layer. Thus it can play on any CD player, even in a car-audio system or a boombox. If you find titles that interest you on hybrid SACDs, you can stock up on them, play them on your existing CD equipment, and get the full benefit of SACD when you buy a compatible player.

Both DVD-Audio and SACD can benefit stereo or surround-encoded music recordings. Neither has been embraced with much enthusiasm by the music industry, which is more concerned with getting control of the file sharing phenomenon. But there continue to be trickles of releases, especially on SACD, from classical labels. You can even add both formats to your system in one box with **universal disc players** that handle DVD-Audio and SACD as well as DVD-Video (and more).

SACD-compatible disc players can output signals in either SACD's native DSD or in high-res PCM. Only a few surround receivers accept DSD but just about all do high-res PCM, allowing users to experience the pleasure of high-res audio.

DVD Plus vs. the DualDisc

It's now possible combine CD audio and DVD features (either DVD-Video or DVD-Audio) on opposite sides of a single disc. The beauty of it is that you can slip it into any CD player to enjoy traditional (i.e. mediocre) CD audio, yet when it hits your DVD player, it can produce video and multi-channel audio. But there's a catch and it's that same old sad refrain—this hybrid CD/DVD is now the object of yet another asinine industry format war.

The first format to emerge in the marketplace was **DVD Plus**. This bonded disc can hold 78 minutes of audio on its CD side and 135 minutes of audio and video on its DVD side. Its inventor, Dieter Diercks of Germany, holds patents on CD/DVD technology in Europe, Japan, Australia, and the U.S. The first DVD Plus title released in the United States was Kathleen Edwards: *Live at the Bowery Ballroom* (Rounder). European labels account for another couple of dozen titles.

And then there's the **DualDisc**. This is another bonded disc, though it holds only 60 minutes on the CD side, and a single layer (about 120 minutes) on the DVD side. It was test-marketed in Boston and Seattle in February 2004 by all five major record labels. In 2005 the format accelerated with releases by Bruce Springsteen, Jennifer Lopez, and other high-profile artists. Releases since then have been scanty.

Both formats have had to overcome a technical hurdle. In order to fit certain CD players—especially slot-loading car players—the discs had to be slimmed down considerably. That in turn may lead to disc-reading problems on some players. My sources say that DVD Plus has done a better job of maintaining compatibility. I've had problems playing the CD layer of DualDiscs—in my opinion, the whole format is a lemon.

Progressive-scan DVD-Video players

Owners of digital television sets should check out **progressive-scan DVD players**, a significant high-end alternative in DVD-Video. These players are able to recognize, and compensate for, the various forms of mayhem that happen to movies as they go from film elements, to master videotape, to the MPEG-2 video compression format at the root of DVD-Video, en route to your DTV screen. Specifically, high-end players deliver pictures to the video display in full frames. In doing so, progressive-scan DVD-Video players provide a more "filmlike" image. (Those who like to avoid complex technical explanations may wish to skip ahead.)

Older DVD-Video players deliver the picture in an interlaced scanning format, known as **480i** (SDTV), intended to suit now-vanishing analog TVs. Progressive-scan players, which now dominate the DVD market, **de-interlace** the picture and deliver it to the display in a **480p** (EDTV) format that better suits digital TVs. (Some go a step further, upconverting to 1080p for 1080p sets.) In addition to delivering a full-frame progressive picture to the video display, progressive-scan players also eliminate any lingering effects of the interlacing and 3/2 pulldown processes that may have crept in prior to or during disc mastering. Therefore a really good progressive-scan player has to perform some fancy footwork to deliver a clean progressive picture to your screen.

The next logical questions are: What is interlacing? What is the 3/2 pulldown? And why is it so important to eliminate all traces of them?

Interlacing is the old analog process that chops up each frame into a pair of separate half-frames called **fields**, with gaps between the scan lines. The video display reunites each pair of fields into a full **frame**, with no gaps between scan lines. Efficiency in use of over-the-air spectrum is the reason why interlacing was invented—it can be thought of as the analog grandparent of today's digital video compression formats. That's why interlaced scanning is a staple of analog television as well as certain DTV formats.

If you're still using an analog TV, don't worry—interlacing is not an issue for you, it's just the way analog video technology works. But the interlacing process is an undesirable element in digital video because it can distort moving objects and camera moves. These interlacing-related video distortions are known as **motion artifacts**. On a big screen, their presence can become painfully obvious not only to golden-eyed videophiles but to anyone with decent eyesight.

The **3/2 pulldown** is necessary to convert film (which runs at 24 frames per second) to video (which runs at 30 frames, or 60 fields, per second). It stretches film's 24 frames across analog video's 60 fields by repeating one field in every other frame. A pair of matching fields alternates with a trio of mixed fields, hence the name 3/2 pulldown.

Alternating normal frames with a jumble of patched-together fields can have messy results. It makes smooth pans look jerky and still frames look flickery. And this frame-rate conversion is totally unnecessary if you're watching a properly mastered DVD (with 24 frames per second) on a DTV (which converts everything it receives to one of its own native formats).

131

In DVD mastering at its best, the 3/2 pulldown is reversed before the film content reaches the MPEG-2 encoder that compresses the video signal for disc mastering. The result should be 24 frames per second on the disc, similar to the original film content, though the frames may be encoded as 48 interlaced fields, arranged in pairs. Each pair represents a moment in time, and each field is marked with a progressive flag that enables a full frame to be reconstructed. Some progressive-scan players use the flags to recognize the paired fields. More sophisticated players can recognize fields without using the flags. In either case, progscan players convert the image to full-frame progressive scanning at 60 frames per second.

In the best of all possible worlds, progressive-scan DVD players would not have to undo the effects of interlacing and the 3/2 pulldown because they wouldn't make it onto the disc. In the real world, these analog quirks live on in expensive telecine and videotape mastering equipment. On top of that, stuff just happens—errors can enter at any stage of the mastering process as the movie goes from film elements to tape master to disc master. A good progressive-scan DVD player has to be nimble enough to detect errors and compensate for them.

So here are a few things that might happen to your favorite movie when you watch it through a progressive-scan DVD player: It starts as film, running at 24 frames per second. The film is transferred to videotape using one of two devices—either a tube-based flying spot scanner or a solid-state CCD scanner. Opinions diverge concerning the aesthetic merits and demerits of these two devices. The chosen device may run at 24 frames per second or 60 fields per second. And the videotape recorder that receives the signal may run at 60 fields per second. Then, if the MPEG-2 encoder receives 24 frames, it encodes them as 24 frames. If it receives 60 fields, they are converted back to 24 frames. If any errors occur during these processes—and they do—then the progressive-scan DVD player will analyze the motley succession of fields and convert it back into full frames ... more or less. (The digital video display performs the final act of conversion, upconverting the progressive-scan signal to its own native scanning rate, but that's a separate subject.)

This is an awesome (and often unpredictable) series of digital video processing events. How the progressive-scan player handles the third phase of it—recognizing and reversing any aftereffects of interlacing and the 3/2 pulldown—is where things get really interesting.

The following is a necessary oversimplification, but in ascending order of quality, a progressive-scan player may: Simply repeat each scanning line of a single field to form a full frame (**line doubling**). Generate a new line, dot by dot, looking at both the previous and next lines, and making an educated guess (**interpolation**). Or generate a new line by looking at individual **pixels** (dotted picture elements) across three or more fields (**field-adaptive de-interlacing**). The latest de-interlacing circuits go a step further, gulping up two whole frames worth of video data to predict how moving objects will change from frame to frame. Slicker ones are good at reproducing mixed film and video content (such as pictures with superimposed lettering) without blurring either part.

Today's best consumer-level progressive-scan players provide a picture with greater vertical and temporal resolution, less flicker, and a nearly invisible line structure that encourages the viewer to sit closer to the screen. The video image has more of the look and feel of film.

While early-generation progressive-scan DVD players came with prohibitive pricetags, the good news is that progressive-scan capability now comes with a pricetag as low as two figures. Anyone who can afford a cheap DVD player can afford a progressive-scan player.

Finally, remember that only a digital television set can benefit from progressive scanning. And while a good progressive-scan DVD player can improve upon a DTV with a bad line doubler, some DTVs have better de-interlacing chips than some DVD players, and therefore may actually look better when fed with the player's interlaced output.

The many faces of DVD for PC

In addition to the **DVD-ROM** (read-only) drives that have spread throughout the PC world, there are several **recordable DVD** formats. In order of seniority, they are:

- DVD-R (1997)
- DVD-RAM (1998)
- DVD-RW (2001)
- DVD+RW (2001)
- DVD-R DL, DVD+R DL (2004)

The **DVD-R** format came first. Its specification was designed by the DVD Forum, a consortium of more than 200 electronics companies. DVD-R and its rewritable companion **DVD-RW** are often designated as the "dash" formats (with a hyphen, not really a dash, in their acronyms). Then Sony, Philips, and Hewlett Packard came along with the "plus" format, **DVD+RW**. Needless to say the dash and plus formats are mutually incompatible. But there are black-box recorders as well as PC drives that handle both, so if you play your cards right, you needn't worry about betting on the loser. **DVD-RAM** has random access capability, allowing it to rewrite portions of a disc, and making it superior for computer data recording. It has made inroads into Panasonic video recorders as well as Hitachi camcorders. The **DVD-R DL** (Sony's name) and **DVD+R DL** (Philips' name) are dual-layer formats that hold up to 8.5GB, enough for four hours of DVD-type (MPEG-2) video. With list pricing dropping well below $300, the DVD recorder has become a viable alternative to the VCR.

Hackers have made it their business and their passion to break the copy-prevention feature that allowed DVD to make its debut amid a brief honeymoon between consumer electronics manufacturers and Hollywood movie studios. The first well known Hollywood-vs.-hackers case involved a copy-prevention-cracking utility called **DeCSS** (CSS stands for Content Scrambling System). As the case was winding its way through the courts, word came that the same decrypting can now be done in seven lines of code.

A related utility confusingly named **DivX** (no relation to the original Divx pay-per-view DVD format) allows decrypted video to be re-encoded in the Windows Media Player, with **MPEG-4** video compression. The resulting efficiency makes it a viable format for file sharing. DivX has been called the MP3 of video and for that reason probably raises Hollywood's hackles. It is included as a feature in a limited number of DVD players. For more information see divx.com.

Connections

- HDMI output
- DVI output
- IEEE 1394-DTCP output
- network connection
- component video output (progressive or interlaced)
- S-video output
- composite video output
- digital outputs (optical or coaxial)
- stereo analog audio line outputs
- 5.1-channel analog audio line outputs

The **HDMI** digital audio/video interface is standard equipment in high-def Blu-ray players and standard-def DVD players. The most desirable versions of HDMI are 1.3 and up if you want lossless audio, or 1.4a for 3D, though any version will feed video to an HDMI-compliant TV as long as both devices support copy protection.

DVI is rare. The 1394 interface, also rare, allows for networking. A few DVD players also come with ethernet connections with may pull content off a router-connected PC's hard drive or stream programming from Netflix, et al.

High-end players also provide **progressive-scan output** through their HDMI or **component video** jacks. They may also have **S-video** and **composite video** outputs for analog TVs.

All Blu-ray and DVD players include **digital outputs** that feed a receiver with Dolby Digital and DTS surround and CD stereo signals. These outputs may be the **optical** or **coaxial** types or both. Note that these do not convey lossless surround at full resolution. Many add extra digital outputs to feed a digital recording device or a digital-to-analog converter. Unfortunately, the digital outs of both DVD-Video and DVD-Audio players deliver a truncated 20-bit/48kHz version of the 24-bit/96kHz signal to allay the music industry's copyright concerns. That's above CD-quality but well below the maximum audio quality level of which DVD-Audio is capable. (Occasionally, with some players and discs, if copyright-protection flags are not in place on the disc, the player will pass a true 24/96 signal. That's the exception, not the rule.)

The **stereo analog audio** outputs usually provide a 2-channel mixdown of surround formats, as well as CD signals converted to analog. The two-channel mixdown may include embedded Dolby Surround, compatible with Dolby Pro-Logic processors in both older and newer surround receivers.

In Blu-ray, the **5.1-channel analog audio line outputs** are valuable for outputting next-generation surround codecs like Dolby TrueHD, DTS-HD Master Audio, and uncompressed PCM to pre-HDMI receivers with 5.1 analog-ins. The surround-capable analog jacks may also be useful in other situations. For example, DTS-HD MA only travels through HDMI 1.3 and up, so if your player and/or receiver lacks HDMI 1.3 and high-res PCM, you'll need the 5.1-channel analog interface to get the high-res signal from player to receiver. Not all Blu-ray players have 5.1 analog-outs, so if you need them, study spec sheets.

HDMI 1.2 will pass SACD signals, and HDMI 1.1 will pass DVD-Audio. Players can also be set to output these formats as high-res PCM, which is compatible with most current receivers. In players without HDMI, a set of analog outs is necessary to feed high-resolution surround signals from a DVD-Audio and/or SACD player to the 5.1-channel inputs on a surround receiver (whether the signal is surround or stereo). A select few universal disc players and receivers support the use of a single digital connection—for example, Pioneer, and Denon Link—but this is regrettably rare. The analog line connections can also be used for Dolby Digital and DTS movie soundtracks. However, if analog line connections bypass the receiver's crossover and bass-management settings, you'll be limited to the bass adjustments provided by the disc player.

Shopping

The most sophisticated shopper will want to put players through a short but revealing series of test scenes. (An old favorite is the opening scene from *Star Trek: Insurrection,* where the camera pans across an urban landscape whose geometric forms easily reveal jagged digital artifacts that result from unsubtle video processing.) In addition to BDs and DVDs, try CDs, as well as anything else you want to play, such as the various recordable DVD formats, CD-Rs, CD-RWs, or MP3s burned onto CD-R/-RW.

How quickly does the machine boot up, load a disc, and respond to disc transport commands (play, scan, etc.)? That will affect the experience of using it. If you're not comfortable with its speed and rhythm, or if the interface makes you work too hard to select and maintain preferred settings, then the first flush of enthusiasm will give way to disappointment, and the machine will seem more an adversary than a helper.

Are you satisfied with the selection and layout of the **front panel controls**? The buttons or joystick used to navigate disc menus are especially important—you'll be using them every time you start viewing a disc. In a nice touch of showmanship, some models sport a motorized swing-down front panel. Does the **remote control** fit your hand, differentiating transport controls by size and shape, placing these and other commonly used functions where your fingers expect to find them? Anything feels right to your fingers is a good thing. In darkened home theaters, backlit or glow-in-the-dark keys can be a big help.

Satellite receivers

Monster dishes that dominated a lawn have almost completely given way to small dish systems that sprout from the roofs and windowsills of both suburban and urban homes. Both of the major small dish systems link an 18-inch dish to a rack-sized receiver box. (Residents of Alaska may need a larger dish.)

DirecTV Satellite System (formerly known as **DSS** until lawyers for a different DSS begged to differ) and **EchoStar**'s **Dish Network** are both small-dish systems that bring anywhere from 100 to more than 200 channels of programming including pay-per-view movies and events. The hardware is generally cheap while the programming is competitive with cable TV rates in most areas. Both are increasing their HD offerings.

Picture and sound quality

Like Blu-ray and DVD, satellite video uses digital compression to efficiently provide a high-quality picture. The quality of that video varies according to atmospheric and other conditions, such as channel-crowding on the satellites.

The word *digital* does not automatically signify superiority. Digital satellite video does not always conform to the standards for over-the-air HDTV, although there are now many bonafide HDTV channels going out over the birds. Picture quality can look marvelous on a good night, terrible on a bad one.

Unfortunately, bad nights have become more frequent. The constant pressure to add new channels to the finite number of satellites in orbit has reduced overall picture quality. New choices are great, but unless they're accompanied by enough new bandwidth, something has to give. A picture filled with blocky-looking digital distortion has become the bane of the keen-eyed satellite viewer. HDTV over-the-air broadcasts generally look better than satellite- or cable-delivered HDTV.

Both DirecTV and the Dish Network offer **Dolby Digital** surround sound on some premium movie channels as well as selected pay-per-view movies and music events. You'll need a Dolby Digital-compatible satellite receiver to get it. If your older satellite receiver is not compatible with Dolby Digital, you can still receive the Dolby Surround signal embedded in the stereo soundtrack and decode it using your surround processor's Dolby Pro-Logic or Pro Logic II modes.

Recording options

If you'd like to use your satellite service with a tapeless **DVR** (digital video recorder) or **PVR** (personal video recorder), both DirecTV and the Dish Network offer satellite receiver packages with built-in hard drives, so you can pause the program while raiding the fridge. High-definition DVR recording is also available from DirecTV/TiVo and the Dish Network.

You can also record from satellite with VHS-family VCRs in the **D-VHS** (the *d* is for *data*) format. (Some D-VHS models record only from satellite, while others record all 18 DTV formats from off-air and other sources.)

HDTV via satellite

Getting high-definition television by satellite requires a different-shaped dish than the conventional 18-incher to receive HDTV. DirecTV's HD broadcasts originate from five different satellites, so they require a satellite dish with five **LNBs** (internal antennas). The newest 5LNB dish has the word Slimline printed on it. You'll also need the HR20 or H20 HD receiver plus a **B-Band Converter** to access new HD channels. DirecTV now supports 1080p video with the right hardware.

DirecTV offers HD versions of channels, where available, for an extra $10 a month over the basic cost of various packages plus $7 for DVR service (well worth it). The selection of packages for sports fans is huge. But when you hear DirecTV make huge claims for HD offerings, be advised that includes satellite radio channels, regional networks not available in all areas, duplicate feeds of certain networks, and other fluff. What most people would regard as the core offerings are more like a couple of dozen channels. Local HD channels are delivered separately by antenna in some areas due to legal restrictions. However, following negotiations with local stations, DirecTV says it now reaches 94 percent of U.S. households with local channels.

DirecTV's HDTV image has fewer pixels than over-the-air HDTV. A 1080i-format HDTV picture normally would have 1920 by 1080 pixels, for a total of 2,073,600 pixels. DirecTV provides fewer horizontal pixels in a 1280 by 1080 picture, totaling 1,382,400 pixels. The picture is not cropped, just lower in resolution, and the difference may not be noticeable on all HDTV displays. The Dish Network's HDTV feed is not truncated.

The Dish Network also offers HD add-ons for $10/month. Local channels no longer cost extra; they are bundled, along with DVRs, in certain packages. The Dish Network's two HD set-top boxes come with a built-in antenna to bring in local off-air DTV stations. These channels also appear in the program guide and can be recorded by the Dish HD DVR Receiver.

DirecTV

This DirecTV 5LNB satellite dish has the five LNBs (or internal antennas) necessary to get HDTV. It is distinguished by the word "Slimline" printed on the dish.

Both major satellite operators market internet access of various kinds through partners. See their websites for more details.

DirecTV and EchoStar (operator of the Dish Network) change their programming lineups and pricing frequently. For the most up-to-date information see directv.com and dishnetwork.com.

Other features

Dual LNB is the name given to satellite receivers that can serve up more than one channel at once. The acronym stands for **low noise block downconverter**. Models with this feature may cost slightly more. You don't need dual LNB to feed more than one TV if all screens display the same signal. Likewise, if you receive a pay-per-view event, and want to feed it to more than one TV, there's no extra charge.

Parental lock can prevent children from viewing adult material.

Connections

- RF satellite input
- RF antenna input
- RF output
- HDMI output
- component video output
- S-video output
- composite video output
- stereo audio outputs
- phone jack
- Dolby Digital output (optical or coaxial)
- data port
- RF remote input

The satellite feed enters your system through an **RF satellite input**. To provide local channels, some of which may be missing from your local satellite feed, there's also an **RF antenna input**. At one time satellite operators were forbidden to provide any local channels, due to federal regulations, but in 1999 Congress began easing these restrictions, following increasing viewer outcry.

The **RF output** will feed older TVs via channels 3 or 4, but the best picture quality will come from the **HDMI** or **component video** outputs. **S-video output** is the next best choice, while **composite video**

140

output is the third.

Stereo audio outputs may carry Dolby Surround, which can be decoded in Dolby Pro-Logic, but a better choice is the **Dolby Digital output**, usually the optical type. Once a relative rarity—early generations of DirecTV (then called DSS) satellite gear didn't have it—Dolby Digital has become standard equipment.

The **phone jack** provided on all satellite gear allows the system to dial into the satellite operator for billing purposes. The **RF remote input** accepts a small antenna that makes it possible to use a remote that penetrates walls with radio frequencies (as opposed to the weaker infrared signals that come from most remotes).

Antennas

Digital broadcast television … public radio and talk radio and Top 40 radio … this is the glorious low-rent entertainment that gushes into our home theater systems via antenna. Over-the-air reception is the first, sometimes the best, and sometimes the only way to get digital TV. It's also the only form of DTV that's free.

While most Americans get television from satellite or cable providers, the antenna is the medium of last resort for outlying areas that aren't wired for cable. Satellite broadcasters are now required to deliver the major networks in the top 20 metro areas, but still do not deliver all local channels in all areas. The FCC adopted "must carry" rules in 2007, ruling that cable systems must deliver all local channels to all sets, digital or analog. But not everyone can get cable. Thus an antenna may be the only way to get DTV with at least some local channels.

As noted in previous chapters, the DTV transition brought analog broadcasting to an end on June 12, 2009. If you depend on over-the-air reception, digital is now the only kind that works. To adapt an analog TV to digital broadcasting, you'll need a set-top box. To feed a DTV with digital signals, you may need a different kind of antenna than the one used previously. That's because your new digital signals may be traveling the airwaves over a different set of frequencies. To find listings of local channels, go to dtv.gov or tvfool.com.

TV channels 2-13 are **VHF** (very high frequency) while channels 14-83 are **UHF** (ultra high frequency). The **lowband** channels, from 2 to 6, are more susceptible to noise than the **highband** channels, 7 to 13. As a result, stations that decided to stay on channels 2 and 6 in the DTV era are having trouble reaching viewers. Channel 6 seems to be especially problematic. Between the lowband and highband channels (6 and 7) is the entire **FM radio** band.

Digital channels may be either VHF or UHF. The DTV transition has caused a game of musical chairs. At the start of the transition, while analog channels were still operating, the Federal Communications Commission assigned unused UHF slots for digital channels. Stations operated at both frequencies for several years, providing both analog and digital service. However, as the analog VHF channels were vacated in 2009, some of the digital channels moved into VHF—enabling networks to reoccupy what they and viewers regard as their traditional slots. The lowband VHF slots, traditionally noisy, are less desirable and some stations are abandoning them for higher frequencies.

Size and range

You're in luck. Smart, well-organized people are trying real hard to make sure you buy the right antenna, with the size and shape that work best in your area. They've provided a color-coded map that'll tell you just what you need. Look for **AntennaWeb** at antennaweb.org, compliments of the Consumer Electronics Association (CEA).

Bigger antennas are for those who live farthest from the transmitter, or who have other reception-strangling problems stemming from local conditions or long cable runs. Those who live in or near major metro areas can get by with a smaller antenna. Bigger is not always better: too much signal may overload the TV tuner. Rather than get hung up on size, focus instead on the antenna's **rated range**, and let the AntennaWeb map be your guide.

Some of the biggest antenna problems are easy to avoid: Splitters should be used sparingly. And attic placement always cuts signal strength by at least half, even in wood-frame construction. A larger antenna might or might not help. With materials other than wood, the problem worsens. Aluminum siding stops signals cold. Other signal inhibitors within your home's walls may include the chicken wire inside stucco, the copper shielding inside terracotta, and brick. If roof mounting isn't practical, consider some other outdoor location.

Local conditions can affect how well an antenna works as much as range. Tall buildings, power lines, lighting rigs, and hilly terrain all may

142

Channel Master, Zenith

Top: The best way to get HDTV is off the air. A Yagi-type antenna like the Channel Master CM 2016 can receive digital channels in both the VHF and UHF bands. Bottom: To adapt old analog sets to the new digital broadcast system, you'll need a digital-to-analog converter box like the Zenith 150-0148 ($69). Analog TV owners received federal subsidies to buy the boxes during the DTV transition.

contribute to reception problems. One of these problems is **multipath distortion**. The term literally describes signals taking multiple paths to the antenna and distorting the picture. When a signal hits a tall building or other obstruction, it splits into pieces. The antenna receives both the main signal and fragments of it that arrive a split-second later. In analog TV, this caused ghosting. Digital TV doesn't have ghosting problems, but multipath distortion still may prevent the set from getting a usable signal. A signal fragment more than 50 percent as strong as the original signal will prevent DTV signals from getting through at all. A more directional antenna will help minimize multipath distortion.

TV antenna types

Outdoor antennas pull in stronger signals than indoor ones, except when mounted in the attic—then they *become* indoor antennas. For outdoor use, choose an antenna based on your reception needs, not cosmetics.

For **indoor** use, you have alternatives, starting with the **dipole rod** and **UHF loop** supplied with smaller TVs, a sculpted designer antenna, or traditional rabbit ears. For FM reception, a designer antenna may look better than a simple (and unsightly) wire antenna, but may not sound better.

Yagi (or **Yagi-Uda**) is the name given the most common type of outdoor TV antenna. Its interconnected rods, or **elements**, are of different lengths to capture the different frequencies that correspond to each channel.

Directional (or **unidirectional**) antennas receive signals from transmitters lying mostly in one direction. This may be a better choice if your local transmitters are all part of the same metro area. Again: consult AntennaWeb.

Multi-directional (or **omnidirectional**) antennas receive signals from all or most directions equally well. This is a better choice if you live between various transmitters.

Motorized antennas include a mechanism that enables the antenna to rotate, facing in different directions to receive channels from transmitters in various locations.

Active antennas boost the signal with built-in amps. Most antennas are **passive**, without amplification, though it can always be added if needed. Just be advised that boosting the signal will not correct defects in the signal (and may overload the tuner).

Satellite dishes sometimes come with built-in antennas to supplement satellite channels with over-the-air channels. There are also antennas designed to mount alongside satellite dishes.

DTV antennas

DTV operates in both the UHF and VHF bands, so antennas that work with those signal types work for DTV. People in most areas will need both. An increasing number of these antennas are being labeled DTV or HDTV antennas.

UHF signals have a shorter wavelength than VHF signals and therefore are captured by shorter rods. It is not unusual for a Yagi antenna to include both the shorter rods that capture UHF signals and the longer rods that capture VHF signals.

Another UHF antenna type is the **bowtie**, whose distinctive triangular rods can be arranged in multiples to pull in signals more effectively. The simplest UHF antenna is the loop supplied with small sets or built into rabbit-ear antennas.

The good news about DTV antennas is that DTV tuners—whether they're separate boxes or built into sets—have seen marked improvement in recent models. That should make any antenna work better.

Radio antennas

FM radio lives in the middle of the VHF-TV band, between channels 6 and 7, so any antenna that works for VHF television will also work for FM radio. To radio lovers, this is an opportunity, but to videophiles it's a problem, so some antennas (and cable systems) will strip out the radio signals with **FM traps** to prevent interference with TV channels 6 and 7.

An FM antenna takes many forms. It may built into the radio, or its power cord, or may be a piece of wire supplied with an audio/video receiver, AM/FM tuner, or table radio. Some wire antennas are Y-shaped to pick up stations in different directions. Some of the most attractive designer antennas are FM antennas.

DVRs, streamers, & servers

Recording devices add convenience to a home theater system. Originally that meant adding a VCR. Today it means adding a DVR, streaming device, or server. DVRs and streaming devices offer ways of capturing content. Servers, on the other hand, exist to organize content, including disc collections.

DVRs

A **DVR** (**digital video recorder** or **PVR** (**personal video recorder**) is a tapeless system that goes well beyond the capabilities of a VCR. If your phone rings, a DVR can pause the action, leaving a still frame while you're occupied. When you return, the DVR can pick up where you left off, or continue the program in real time, or go back to the beginning while still recording the remainder. Once you get used to this kind of convenience, the VCR begins to seem primitive and clumsy. While there are DVD-based and VHS-based digital recorders, the terms DVR and PVR usually refer to hard-drive-based components.

Programmed recording

Recording capacity is a function of how big the unit's hard drive is. Cable systems are just starting to implement **RS-DVRs**, or **remote-storage DVRs**, which replace local hard drives with server storage. Hollywood and the networks tied up this cost-saving idea in the courts but it finally broke free.

DVRs use video compression, as do digital video formats and satellite systems, but there's a tradeoff between the **quantity and quality** of recording—as running time goes up, picture quality goes down. A bad DVR picture looks different than a bad analog VCR picture; rather than going fuzzy, it pixellates, dissolving into an expanse of squarish blocks.

Recording with a DVR is easy thanks to the **onscreen program guide**. Read through it, click on the items you want, and the machine records your selections. There's no need to fuss with start/stop times or type numeric codes. VCRs can have onscreen program guides too, but with a DVR, everything is easier, once you know the ropes.

One recent development in hard-drive video recording is the ability to record the sharper picture available in the **high-definition TV** formats. Some DVRs still record only standard-definition TV—with about the same resolution as a DVD, or less, depending on what recording-time/quality option you choose—but HD-capable products are worth seeking out. DVRs supplied by most video providers (cable, satellite, telco) also support video on demand.

Ways to get a DVR into your system

There are a lot of ways to get a DVR into your system. Hard drives are cheap, so a handful of technically literate home theater buffs prefer to build their own PC-based DVRs with generic parts and ATSC tuners. However, most DVR fanatics prefer an easier route.

If you are a satellite subscriber, check out satellite receivers with integrated DVRs. Likewise, if you're addicted to cable, ask your cable operator how much more per month you'd have to pay for a DVR-equipped cable box. You can also buy component DVRs. If you want your DVR to have a more sophisticated and helpful user interface, check out TiVo.

TiVo: the deluxe DVR

The first generation of DVRs included three branded standalone products that were not limited to a single cable or satellite service: **TiVo**, **ReplayTV**, and **UltimateTV**. The latter two have since retreated under new ownerships, leaving TiVo as the class act of the field. UltimateTV is now owned by Microsoft and marketed as an extra-cost option through DirecTV. ReplayTV left the self-branded hardware realm, evolved into PC software, and is now owned by DirecTV. Their program guide subscriptions continue.

TiVo still sells standalone DVRs but also has arrangements with cable (Comcast, Cox, RCN) and satelite (DirecTV) companies. For the standalone version, TiVo charges a **subscription** fees for its program guide. Fees are $12.95/month, $129 for a one-year contract, $299 year for three years, or $399 for lifetime service.

If you want to avoid subscription fees, several manufacturers offer DVRs with a **PSIP**-based program guide—that stands for Program and

System Information Protocol and it's a data signal transmitted by broadcasters on their on DTV channels. A few DVRs come with built-in DVD players or recorders.

The big news for TiVo fans is a pair of new DVRs: the TiVo Premiere ($300) and Premiere XL ($500). They include a "reinvented" HD interface, full 1080p support, web programming, and streaming via Netflix, Amazon, Blockbuster, and YouTube.

Among older models, the Series3 was the first TiVo to be HD-capable, and CableCARD-compatible (with two slots). It's been succeeded by TiVO HD and TiVo HD XL, the latter with seven times the capacity plus THX certification. TiVo Series2 was first to accept a high-speed network connection with associated features.

TiVo users love to tinker with their systems. You can, for instance, add another hard drive if you know what you're doing. For more possibilities, Google "tivo hacks."

TiVO features

Following are features common to all TiVos from Series2 on:

- basic DVR
- 2-week program guide
- records favorite series, skips repeats (Season Pass)
- searches by title, keyword, performer, etc.
- skips reruns, if requested
- resolves recording conflicts
- remote recording for internet or AOL users
- broadband compatible
- indexes recorded shows alphabetically or chronologically
- folder support for indexed shows
- previews upcoming programs
- schedules recording during previews
- 3 scan speeds (3X, 18X, 60X)
- learns viewing preferences, switchable (TiVo Suggestions)
- TiVolution magazine
- watch recordings on another TV
- online/mobile scheduling of recordings
- transfer recordings to mobile devices
- stream Blockbuster On Demand
- Amazon VOD movies and shows

148

- subscribe to free web videos
- watch free music videos on demand
- access streaming music services
- listen to PC/Mac music library
- view photos on big screen

Following are features added by TiVO HD and Series3:

- recording time up to 20-32 hours HD
- search broadcast TV and internet
- expandable storage space
- supports over-the-air digital channels
- world's largest on-demand video store
- Netflix movie and TV streaming
- YouTube streaming

Following are features exclusive to TiVo Premiere and Premiere XL:

- recording time up to 45-150 hours HD
- new HD interface
- full 1080p support
- browse by TV and movie posters
- find shows based on topical Collections
- see similar shows in the Discovery Bar
- live video window in menus
- 30-second scan
- disc space meter
- Energy Star compliant

Connections for TiVo and other devices

- HDMI, DVI inputs
- 1394 digital in/out
- CableCARD slot
- RF antenna/cable inputs
- component, S-, composite video outputs
- stereo analog line outputs
- RJ-11 phone jack
- ethernet, wi-fi network connections
- USB port (TiVo)

Some (though not all) DVRs come with high-quality digital HDMI, DVI, or 1394 outputs. Some HD recorders also have 1394 input. All formats come with S-video and composite video jacks, plus analog line outputs, which provide stereo or Dolby Surround sound. Also common to all systems are the RJ-11 phone jack, which allows each system to update its electronic program guide, and the IR blaster, which controls your cable or satellite box. The exceptions are the broadband models which require a high-speed network connection. The USB ports added to newer TiVo products allows access to digital photo and music files through a home network. They connect to various adapters.

Movie streaming and download services

DVRs grab programming from over the air, satellite, or cable. But you can also get movies through a broadband internet connection. Some services let you **stream** (or rent) movies or TV episodes while others let you **download** (or own) them.

There is some overlap between the two types of service. For instance, both Apple's iTunes Store and Amazon offer both streaming and download of movies. As this book was going to press, Amazon offered *The Wizard of Oz* for $2.99, streaming; or $9.99, downloading. Content streamed or downloaded from Amazon can be viewed on Panasonic, Samsung, and Sony TVs as well as TiVo and Roku set-top boxes. Content streamed or downloaded from iTunes can play on a variety of Apple devices such as the iPhone and iPod touch.

Retail chains that once just rented movies on disc now allow them to be streamed. The two biggest players in this field are **Netflix**, which allows streaming as an added benefit of the monthly membership fee, and **Blockbuster On Demand**, with a la carte pricing.

Another major player in streaming is **Roxio CinemaNow**, which operates in PCs, TVs, set-top boxes, Blu-ray players, smart phones, and other media devices. Best Buy has licensed Roxio's technology, allowing consumers who buy the chain's Insignia-brand TVs and Blu-ray players to set up the CinemaNow service at the store. Covering all bases, Insignia products also stream Netflix.

Cable and satellite companies offer movie streaming through their **video on demand** services. Subscribers who pay an additional fee per program can access it through a set-top box or authorized DVR.

In the future, the line between streaming and downloading may blur as the entertainment industry explores **cloud-based content services**.

These would let you pay a fee to purchase rights to enjoy a piece of content, accessing it from a variety of broadband-connected devices such as TVs, Blu-ray players, PCs, gaming consoles, smart phones, and tablet computers. One major cloud-based initiative is **UltraViolet**, which is supported by several major movie studios, TV makers, cable systems, retailers, and others.

Music streaming services

The largest source of free audio streaming is **internet radio**, which can be accessed via PC, server, a/v receiver, or standalone radios. Radio Denmark alone has three classical channels. There are also services that aggregate internet radio stations with a convenient user interface. They include **Pandora, Slacker,** and **SHOUTcast.**

If you don't mind paying, there are also **music subscription services** that operate online. You can access them via PC, mobile devices, or some models of broadband-connected surround receivers and Blu-ray players. They include **Rhapsody** and **Napster.**

Satellite radio services from **Sirius** and XM offer a cornucopia of music and talk for subscriptions ranging from $2.99/month for XM Online to $16.99/month for full XM listening privileges. Sirius charges from $6.99/month for a la carte service to $12.99/month for the full Sirius range. Sirius and XM have merged but still operate independently. You can buy their combined services for $19.99/month.

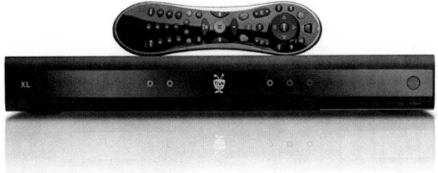

TiVo

The TiVo Premiere DVR (starting at $299, or $499 for the XL version) can record over-the-air HDTV or from cable thanks to its CableCARD slot. This broadband-enabled DVR can also stream video or music from the internet and access media from your PC or Mac.

Music download services

File sharing started the downloadable-music phenomenon but the next step in its evolution is legitimate purchase—that is, music-industry-approved downloads.

The most successful example is **iTunes**, Apple's proprietary service for iPod owners. Other major sources of music downloads are **Amazon** and **Walmart**. For audiophiles, the most significant download service is **musicgiants.com**, which offers lossless downloads in the WMA format at data rates up to 1100 kilobits per second, way more than conventional downloads.

Be warned that some of these services impose restrictions on the use of downloaded material. Most notably, some downloads will work only with some devices. The **iPod** and iTunes are the most notorious example, using a form of the **AAC** file format protected by Apple's **Fairplay DRM** (digital rights management). Many of the other services use Microsoft's **WMA** (Windows Media Audio) with **PlaysForSure** DRM, now known as **Certified for Windows Vista**. Microsoft's new **Zune** player uses a yet another, newly developed DRM. The hazards of DRM have become apparent as some services (Microsoft, Yahoo Music) have gone out of business—and their encryption keys expired, leaving legal purchasers with tracks that couldn't be played or could not be added to new computers and other devices.

However, DRM-free downloads are now hitting the iTunes, Amazon, and Walmart sites, among others. In the case of iTunes, they cost more, and still use the Steve Jobs-favored AAC format (without the DRM). However, many other download services are now selling DRM-free MP3s. This is a great step forward for online music retailing.

Don't neglect the musician-run sites. A lot of up-and-coming bands offer free tunes, and if you want to support the band, you can always buy a T-shirt. **Social networking sites** (MySpace, Facebook) have become another significant source of music for young listeners.

Digital media and audio servers

The term "server" usually refers to a powerful computer at the heart of a network. That's why the term is crossing over into home theater—a **digital audio server** (**DAS**) is basically a computer with audio outputs.

While there are also **digital media servers** that handle both video and audio—**Kaleidescape** is an outstanding example—copyright issues surrounding the copying of video from disc to server have kept them tied up in court. So this section will concentrate more on digital audio servers.

Compressed audio files can enter a home theater system in several ways. Get a digital audio server, a network-enabled surround receiver, a receiver with USB input, an MP3/WMA-compatible BD/DVD/CD player, an iPod dock, or have a RadioShack dongle perpetually hanging out of your receiver's back panel to service a cute little portable player. The networked receiver and MP3/WMA-compatible disc player are the least invasive solutions, but not all receivers have networking, and the disc player limits your audio file consumption to files burned onto CD-Rs and CD-RWs. A digital audio server offers more ways to get MP3 files into your system, not just reading them off a disc, but helping you organize them too.

Capacity dictates how much material you can load into a storage device. A flash-memory player can hold up to 32GB, and hard-drive-based players top out at 160GB, but the better servers holds hundreds of gigs or even tip over into terabyte territory.

Most digital audio servers contain a **CD-R/-RW drive**. It can be used either to load music into the system (from CD to hard drive) or to burn discs (from hard drive to CD-R or CD-RW). A few models dispense with the built-in CD drive but are able to connect to (and control) CD jukeboxes containing hundreds of CDs.

By connecting to the internet you can get **streaming audio** from hundreds of internet radio outlets (many of which carry the signals of regular radio stations).

Products with **upgradable firmware** can be updated to provide compatibility with new file formats.

MP3 and other compressed audio file formats

- lossy vs. lossless formats
- MP3
- AAC
- WMA
- RealAudio
- ATRAC
- OGG (Vorbis, FLAC)
- WAV / AIFF / ALC

Audio servers (and file sharing) require digital signals to be encoded in file formats called **codecs**—that stands for encode/decode. **Lossy** formats uses heavy compression based on **perceptual coding** to eliminate some audio data that are masked by louder, more dominant audio data. These formats omit 80 percent or more of the original data. **Lossless** formats eliminate far less, roughly 50 percent. They do not rely on perceptual coding and are usually higher in quality.

MP3, a lossy format, is the most widely used one. Where did it come from? Originally named **MPEG-1, Layer 3**, it was intended to compress audio on Video CD, a pre-DVD format. Of course, Napster and other file-sharing services subsequently gave it a new lease on life and made it famous. The inventor was the Fraunhofer Institute of Germany. Fraunhofer is a co-licensor of the format along with Thomson (former owner of the RCA and RCA Scenium television brandnames).

MP3 has plenty of company. Other lossy codecs include Dolby's **AAC** (Advanced Audio Coding), the dominant format at Apple's heavily hyped **iTunes** online music store. Microsoft's **WMA** (Windows Media Audio) often goes head to head on streaming-media websites with the **RealPlayer** from RealNetworks. An early use of compressed audio was in MiniDiscs with Sony's **ATRAC** (Adaptive TRansform Acoustic Coding), which helps the MD achieve near-CD-quality recording and playback on a 2.5-inch caddy-enclosed disc that fits into portables, components, and compact systems.

From the open-source community comes a group of patent-free codecs in the Ogg family. One of these is **Ogg Vorbis**, reputed to be among the best-sounding lossy codecs. The Ogg family also includes **FLAC**, or **Free Lossless Audio Codec**. The Mac platform includes **ALE** (Apple Lossless Encoder). The latter is associated with iTunes, QuickTime, and the Airport Express wireless distribution device.

In addition to lossy and lossless formats, there are **uncompressed audio formats** that use no compression whatever. Uncompressed CD audio is called **PCM** (**pulse code modulation**). The main uncompressed format in Windows is **WAV** (WAVEform audio). Apple's equivalent is **AIFF** (Audio Interchange File Format). All of these are equivalent to CD audio. But because they are less efficient than lossy and lossless codecs, they are rarely used in audio servers.

Despite all that competition—and a clear preference among audiophiles for lossless formats—it is the MP3 format that remains virtually synonymous with downloading and ripping. It is MP3 that continues to dominate **file sharing** services serving as havens for the shattered Napster community. Increasingly it also dominates legal download stores.

Other members of the MP3 family

MP3 VBR is a slightly improved version of MP3. Conventional MP3 operates at a constant bit rate. MP3 VBR operates at a variable bit rate, depending on the demands of the music. That allows it to provide slightly better quality in the same number of bits.

A more thoroughly improved version of MP3, **mp3PRO** provides better sound at lower bitrates than the original (and highly popular) MP3 format. It accomplishes this by providing a second stream of digital information. When mp3PRO files are played on an original MP3 player, the older player can access just one of those two streams. Therefore only an mp3PRO player affords the new and improved sound. If you're interested in mp3PRO, upgrading to better PC/Mac software is the easy part. (Thomson's rca.com was the first website to offer the PC-software player.) The hard part is to find mp3PRO-compatible hardware.

MP3 Surround adapts the popular stereo file sharing format to 5.1 channels of surround sound. It's backward compatible with existing MP3.

These two technologies really work—I have heard convincing demos of both—but sadly, they have yet to find widespread applications, and the music industry has shown no sign of interest in them.

Ripping

Of course you can always use your PC and an MP3-encoding program to encode your own MP3s from CDs or even analog sources. (Lots of my MP3s have LP surface noise!) The easiest way to encode MP3 is with **Windows Media 10** and beyond, which rips at blinding speed, and does it without a plug-in. Of course, **iTunes** also does an excellent job. Two other leading encoding programs are **WinAmp** (winamp.com) and **MusicMatch** (musicmatch.com). Both are available in no-frills basic or full-featured pay versions and both also serve as software-based players.

Quality is an issue in **MP3 encoding** (and collecting). Like Dolby Digital and MD/ATRAC, MP3 uses compression. None of these provides a CD-quality (uncompressed or lossless) signal. The bare minimum for MP3s is 128kbps (kilobits per second), though it doesn't sound very good. You'll get better results at 192kbps, and 320kbps approaches CD quality in most portable or casual listening applications. Some other encoders, like AAC and WMA, achieve higher quality at lower bit rates.

To rip from vinyl, you'll need a **phono preamp** with USB output to feed your PC. NAD, Pro-ject, and Bellari make inexpensive and good-sounding USB phono preamps. Another method would be to burn the vinyl to CD-R/-RW and rip the latter disc on a PC. To edit audio tracks on a PC, Audacity is an excellent choice, both versatile and free.

Metadata

Metadata are data about data. When your music player or server sorts material by artist, album, song, genre, etc., it's using metadata. **Gracenote**—once an open-source project, now a profit-making company—is the largest provider of metadata, and the favored provider for iTunes, but other metadata lookup services work with various products. The Windows Media Player works with **AMG** (the All Music Guide) and many other providers. **Musicbrainz** (musicbrainz.org) is a surviving open-source database used in many PC applications.

When you rip from CDs with iTunes, WMP, or another encoding program, the metadata may come up automatically. Or you may be asked to confirm which album you're ripping. Older versions of WMP provide many choices for metadata lookup.

If you rip from cloned CDs or analog sources, you may need to edit the metadata to ensure that your player or server can find material by artist name, title, etc. You can do this manually, through iTunes, WMP, or even the Windows Explorer. But there may be an easier way. WMP will do an info search that lets you specify artist or album, find the metadata, and associate it with the track before or after ripping. Gracenote can identify a cloned CD using audio fingerprinting.

The Gracenote database and services

Digital audio servers that connect to the internet offer some significant convenience features. Many of them are provided by **Gracenote**. Now owned by Sony, this is the company that provides iTunes and the Creative Labs Zen with metadata.

The Gracenote music recognition database (formerly **CDDB**) aggregates track metadata, 24 hours a day, from 130 million users all over the world. It has downloadable data on millions of CD titles and tens of millions of song titles.

Gracenote's **MusicID** uses a fingerprinting technique to recognize not only whole CDs but individual songs in any format (MP3, WMA, AAC, etc.).

Gracenote **Mobile Music ID** identifies songs via cell phone. Hold your phone up to the radio and the database identifies the track. Then you can purchase the music as either a download or a ring tone.

Gracenote **Playlist** works in PC, mobile, and car products. Select an artist or track as a "seed," and it generates and updates playlists dynamically using Gracenote's music similarity data and global popularity trends.

Gracenote **Playlist Plus** does the same for flash-based devices that may lack internet connectivity or memory.

Gracenote **Link** aggregates and delivers third-party content while music plays.

Gracenote **Music Enrichment** combines the sonic excellence of mp3PRO with the music recognition database, providing artist and title info, bios, photos, links, discographies, artist news, tour info, lyrics, and recommendations.

Finally, Gracenote **VideoID** organizes Blu-ray and DVD collections (in a changer or server) by title, release date, language, regional code, single/double-layer info, credits, running time, color/B&W, aspect ratio, and audio formats.

Copyright and copywrong

What the music industry calls **copy protection** is a hot issue. Some anti-copying technologies are designed to prevent CD-R copying though some also prevent MP3 encoding or transfer. Some of the more insidious ones prevent CD-ROM drives, the kind found on a PC, from accessing CD audio data. Such CDs cannot be encoded into MP3 form because they cannot be read by the PC at all. It would take another book to describe each anti-copying system and list the affected releases, but the two most notorious examples are **XCP** and **Mediamax**, which landed Sony BMG in lawsuit hell. These two nightmare technologies installed themselves on PCs and exposed them to security risks. The record label was forced to withdraw the tainted CDs and replace them on request with untainted ones.

157

Another key use of copy protection is to enforce the binding of hardware and software. For instance, older iTunes downloads in the FairPlay-protected AAC codec can be played only on the iPod (no other players) or on a computer. Several other download services and players use Microsoft's WMA and its various DRM schemes. However, if you really want to cross these proprietary boundaries, hackers offer many transcoding programs. Just be sure anything you download for this purpose doesn't come loaded with spyware.

Perhaps the hottest issue of all is the fact that the Recording Industry Association of America, the U.S. music industry's trade association, filed nearly 30,000 lawsuits against listeners who engaged in file sharing. This was the most brilliant idea the industry came up with since EMI paid Mariah Carey $21 million not to make any more records. Why bother with improving your download pricing, selection, or sound quality when you can just litigate against your own potential customers?

More recently, the RIAA and its European counterparts have been jawboning internet service providers to issue warnings—and eventually cut off service—to persistent illegal downloaders. Though not unfair in my view, this strategy will hinge on the cooperation of ISPs, which so far has been hit and miss.

The music industry's legal campaign got a leg up with the Supreme Court's decision in MGM vs. Grokster. Unanimously ruling in favor of the music industry and against the file-sharing service, the court did not say outright that file sharing technology itself is illegal. Rather, it criminalized the use of technology with *intent* to infringe the copyright law. The inevitable result of this non-decision will be more lawsuits.

The RIAA and the Consumer Electronics Association wage an ongoing battle against and for (respectively) the consumer's right to record for personal use. Feel free to let your elected representatives know where you stand.

Connections for servers and home networking

- digital audio input/output (optical or coaxial)
- stereo analog line input/output
- HDMI
- headphone output
- FireWire port
- USB port
- network or phone connection

- wireless connection
- powerline connection
- iPod docking connector

Digital audio servers connect to your system like any other digital component, using digital optical or coaxial connections, stereo analog line connections, and (in media servers) HDMI and various other video connections. A PC connection, especially ethernet, but also FireWire or USB, is another likely addition, but systems vary.

The most deluxe servers are designed to work with PCs and **networked audio** systems. The latter are easy to arrange in new home construction; retrofitting an existing home is harder but not impossible. The most common kind of home network cable uses **internet protocols** and a **router** (to split and distribute the signal). The wiring is usually **Category 5** with **RJ-45** connectors. Cat5 cable carries the equivalent of four separate **twisted pair** (phone) lines though it can distribute various kinds of signal. A variation called **Cat5e** has a center spine that separates the four twisted pairs, reducing crosstalk, and therefore allowing greater bandwidth which translates into higher-speed data transfer.

However, especially in older homes where new wiring is an intimidating task, there are alternatives. Some products now support **wireless** signal-transfer schemes such as the **802.11b** standard (better known as **Wi-Fi**). You can see Wi-Fi enthusiasts huddled over their laptops in coffeeshops and airports. The *b* version operates at a pokey 11 megabits per second—plenty for MP3 but not for video—and is used by Apple's AirPort Express. **802.11a** is five times faster, at 54Mbps, but is not compatible with the *b* version. However, **802.11g** operates at the same speed as *a*, and is backward-compatible. What's next? That's **802.11n**, which transmits up to 540Mbps at distances up to 160 feet. What's beyond that? Possibly **802.16**, a.k.a. **WiMAX**, a broadband wireless access standard which extends range from 300 feet to several miles. It may someday replace wired broadband conections (or at least provide some stiff competition). A wireless version of **IEEE 1394**, also known as FireWire, was finalized in 2004 and offers an attractive blend of high bandwidth and convenience.

Products with the **Home PNA** standard use **RJ-11** phone jacks and existing phone wiring to spread networked audio throughout a home. The phone line can be used for calls at the same time as long as the line isn't being used for a dial-up net connection. Home PNA 2.0 has a maximum speed of 10 megabits per second (more than enough for MP3 data rates). Just finalized is a **Home PNA 3.1** standard that would increase the data rate over phone wiring to a speedy 320Mbps over two simultaneous channels, enough for compressed video. For more information on the Home Phoneline Networking Alliance visit homepna.org.

Another non-invasive method of networking called **HomePlug** uses existing home power lines for audio, video, VOIP phone, and home networking control. The latest version, **HomePlug AV2,** has improved throughput speed, transmission range, and operating spectrum. You can use it to network a broadband connection, multiple 1080p video streams, 3D video, 2K by 4K ultra-res HDTV, security cameras, and internet gaming. For more details on the HomePlug Powerline Alliance see homeplug.org.

Finally, the iPod has become an audio server in itself. The universal popularity of Apple's iPod has birthed a whole cottage industry of iPod-connected products. They use the 30-pin **iPod docking connector** built into most iPods (except the Shuffle). Some of these devices are docks that connect to surround receivers, compact systems, or even multi-zone audio systems. Some are standalone systems in themselves, with their own drivers and amps, that use the iPod as the key (though not necessarily the only) signal source. A select few docking devices offer a high-quality digital connection. Others use an analog line connection, which still sounds better than the player's headphone output. In addition, an iPod can connected to some surround receivers using a direct USB input with the Apple-supplied cable. However, any music player (including the iPod) can connect to a system using its **headphone output**, which is basically a volume-controlled analog stereo line output. You'll need to get an adapter cable with a 1/8-inch mini-plug at one end and two RCA-type plugs at the other. Manufacturers like Monster and Tributaries make very good ones; if you're desperate, there's always RadioShack.

VCRs

Though it has joined the ranks of "legacy" components, and was never a stellar performer, a stereo VCR (videocassette recorder) is a must for anyone with a library of videotapes. It provides low-cost recording with removable storage and, sometimes, access to stereo or Dolby Surround soundtracks.

Some VCRs are better than others. While any VCR will play movies, only D-VHS will record HDTV, thanks to its ATSC tuner (plain-VHS tuners are the obsolete NTSC type). Other step-up options include Super VHS (best experienced in ET, its latest extension), electronic program guides, editing features, and the hard-to-find manual recording-level control.

VCRs, home theater, and history

Ronald Reagan was president when the "video revolution" turned television into something that could be time-shifted and recent Hollywood movies into something that could be rented. Livingrooms became screening rooms and home theater was born.

Consumers had a choice of formats, and while Sony's **Beta** format lost, it left behind the 1984 Betamax Decision, in which the Supreme Court ruled that home video recording for personal use is not illegal. Sony's Beta hi-fi stereo and SuperBeta innovations also preceded similar improvements in VHS.

But the format war's victor was **VHS (Video Home System)**, from Matsushita's JVC subsidiary, the first format to accommodate the two-hour-plus running times that make movies watchable in a single big gulp. Why dwell on this historical material? If it has any meaning, it's that cinematic values—not just higher performance—ruled the day.

VCR formats

VCR or DVR? Either a traditional videocassette recorder or a tapeless digital video recorder can time-shift programming to a more convenient viewing time. DVRs (also known as personal video recorders, or PVRs) offer niceties that go beyond traditional VCR features, such as the ability to pause the action during real-time viewing. But only a VCR will play your old tapes and provide removable storage using the universally accepted VHS tape. Some manufacturers offer combination players containing VCR/DVR or VCR/DVD in a single chassis.

Super VHS offers 50 percent more picture detail than regular VHS. S-VHS is backward-compatible in that it readily records and plays standard VHS tapes. Unfortunately, the reverse is not true: S-VHS tapes will not play in most regular VHS VCRs, except for those with **quasi-Super-VHS playback**. While S-VHS looks sharper than VHS, it does

161

not look any cleaner, because its superior method of recording applies only to the brightness (luminance) portion of the video signal, not to the color (chrominance) portion.

The latest version of S-VHS sports a couple of extra letters. **S-VHS ET**—that's Super VHS with Expansion Technology—records in S-VHS using regular VHS tape. The tape must be high-grade but needn't be the special, expensive tape previously required by S-VHS. VCRs with the aforementioned quasi-S-VHS playback can play S-VHS tapes at lower standard-VHS resolution, but cannot record S-VHS tapes.

D-VHS (*d* is for *data*) is the latest addition to the VHS family. There are two kinds of D-VHS. Originally the format was designed to record only from satellite systems, and some D-VHS models still have that limitation. But other models can record in any of the 18 ATSC-sanctioned DTV standards. That makes D-VHS potentially controversial—Hollywood frowns on digital home video recording. To address that concern, the DTCP variation of the IEEE 1394 interface includes a trio of copy prevention features: copy freely, copy once, or copy never. (For more details about DTCP see "Television/DTV connections.") In addition, VHS-family format inventor JVC introduced an extra layer of copy prevention called **D-Theater** in an attempt to persuade Hollywood to release 1080i HD movies in D-VHS. A handful of movies were released but this idea never really took hold.

The word digital can be used misleadingly. Be warned that a VCR with **digital effects** is not necessarily a digital VCR. It just uses digital tech in the form of a computer chip that stores the image for noiseless freeze frames and other effects. Similarly, the phrase **digital noise reduction** has nothing to do with DTV. Here digital circuitry is being used to reduce video noise. Don't confuse VCRs with digital effects or digital noise reduction or digital anything else with true digital VCRs.

Standard features

Most old VCRs have a now-useless analog **TV tuner** (channel selector) and a **TV/VCR switch** to select between tuner and tape playback. Most (though not all) come with a **front panel display** and **wireless remote control** (which may be preprogrammed to control other devices). Basic **transport controls** (play, pause, etc.) are always provided, though the quality of **special effects** modes (picture scan, slow motion, still frame) will vary depending on the number and type of recording/playback heads provided.

The two standard recording/playback speeds are **SP** (**standard play**) and **EP** (**extended play**), also known as **SLP** (**super long play**). SP, the fast speed, records 120 minutes on a standard T-120 cassette, or 160 minutes on a T-160. EP, the slow speed, records six hours of material on a T-120 or eight hours on a T-160. Some older machines have a four-hour **LP** (**long play**) speed for recording and/or playback. However, the LP speed is now an anachronism, not because it shares a name with the 33rpm long-playing record, but because it eats up more tape to record a picture of about the same (poor) quality as EP.

Automatic features

Numerous automatic features make newer VCRs a little slicker than early generations. Push a cassette into the slot and **auto on** powers up the machine—no need to hit the power button. **Auto rewind** shuttles the tape back to the beginning when the show is over.

In VHS hi-fi stereo models, **auto recording level control** ensures that audio levels are evenly recorded. Users sophisticated enough to want **manual recording level control** should start looking at editing decks.

Auto head cleaning wipes dirt off the rotating recording/playback heads when you insert a tape. Better blank videocassettes also perform head cleaning when in use. All that's good, because dirt can do damage to both VCR and tape.

Now automated on most decks, the **tracking control** eliminates onscreen distortion by compensating for slight differences in the positioning of the recording/playback heads on different VCRs.

Convenience features

Time-shift recording, a means of moving programming to a more convenient viewing time, requires you to learn how to program the **tuner/timer**. Odds are you won't be using a VCR for this anymore, unless you have a D-VHS machine, or are recording analog channels in a cable system.

Time-shifting was never as daunting as late-night comedians claim. Modern VCRs tape at least eight programs (**events**) over a limited period (as much as a year). Two advances have made time-shifting easier. **Onscreen programming** replaces the tiny front-panel buttons of 1980s VCRs with step-by-step instructions on the TV screen.

If that isn't easy enough, you may prefer the second advance in VCR programming, found more often in step-up models. **VCR Plus+** (nono, that's not a typotypo) was originally a separate device that resembled a remote control, but it has become a common step-up VCR feature. VCR Plus+ sets the clock and controls both your VCR and your cable box to make time-shifting even easier. You can set your VCR to record programs by punching in numeric VCR Plus+ codes published in most TV listings. Related options include **Guide Plus+**, with electronic program guide, and **Index Plus+**, which stores program title, length, and an ID number on a VCR memory chip in order to track up to 400 recorded programs.

Commercial advance blocks ads from appearing on recordings, replacing them with a blue screen. **Movie advance** automatically fast-forwards through trailers and ads at the start of a rental tape. (Older folks like trailers because they remind us of the butter-reeking, popcorn-strewn moviehouses of our youth.) **High-speed rewind** is another plus. Better VCRs actually vary the rewind speed, slowing down when they near the beginning of the videocassette.

For those actively engaged viewers who like zipping back and forth to examine a sequence in variable-speed fast scan or frame-by-frame slow motion, a **jog/shuttle dial**, at deck or remote, may be preferable to the traditional button-punching interface.

Sound features

For a home theater system, **VHS hi-fi stereo** recording and play-back capability is a must to extract Dolby Surround sound from stereo movie soundtracks. This audio-for-video system records high-quality analog stereo signals along with picture information in a process impressively known as **depth multiplexing** (no, that won't be on the spot quiz or the spec sheet).

VHS hi-fi is far superior to the other soundtrack carried by most VHS videotapes—a thin sliver of mono that sits on the tape's fragile edge. Once used for a low-grade form of stereo, only slightly improved by Dolby B noise reduction, that **linear** soundtrack is used today to allow mono tapes to play on hi-fi and mono VCRs alike. But only a hi-fi VCR and tape can deliver stereo and surround sound, which are not luxuries, but necessities, in a home theater system.

Any model that records and plays hi-fi stereo will have **MTS (multichannel television sound)**. Back in the heyday of analog broadcasting, that would allow stereo to be recorded off-air. It may still find use with analog cable channels. Hi-fi recording and MTS reception are separate features but usually go hand in hand. All MTS tuners have **SAP** (**second audio program**) capability for bilingual and other uses.

Are more heads better?

Any VCR comes with at least two rotating heads for recording and playback. In analog VCR recording, each head records one field, or half a frame, skipping every other scanning line, and the fields interlace into a full frame to form the picture. A **two-head** machine is thus a basic one.

Four-head VCRs come with a second pair of heads optimized to provide higher quality in special effects modes (picture scan, slow motion, still frame). These step-up VCRs are generally better built, more fully featured, and more suitable for frame-by-frame video editing. Special effects may also be provided (even in analog VCRs) by digital circuitry. Add one more pair of heads for hi-fi stereo recording and the VCR may be described as a **six-head** model.

Most attractive are VCRs with **19-micron** (read: narrower) heads, which greatly improve recordings made at the EP speed, narrowing the performance gap between slow-speed and fast-speed recording. This feature is often combined with **flying preamp** circuitry, in which video amplification is moved to the spinning head drum, reducing video noise.

Editing deck features

No one edits on tape nowadays. But back in the day, the following features were used.

A real editing deck will have **edit-control jacks**, which synchronize playback and recording functions in two decks, and **front-panel inputs** to easily accept input from a camcorder.

A **flying erase head** enables clean edits by removing noise bars (streaks and lines) between recorded scenes or segments. It also makes slow and fast scan modes look better.

To provide videomakers with additional control over audio, the **audio dub** feature adds a soundtrack or voiceover narration to existing video. It affects only the mono linear soundtrack. **Manual recording level control**, combined with volume meters, affords more control over audio levels, which are adjusted automatically in most VCRs.

Connections

- RF input/output
- composite video input/output
- S-video input/output (Super VHS VCRs only)
- IEEE 1394 (FireWire) input/output
- HDMI output
- component video output
- stereo analog line audio inputs/outputs

All VCRs have an **RF input** to accept an antenna, satellite, or unscrambled cable feed. There's also an **RF output** to feed an older TV (the kind with only an antenna input). If your ancient TV has a 75 Ohm antenna input, just plug the RF cable right into it. If your museum piece has a 300 Ohm antenna input, you'll need a simple "balun" adapter. Either it came supplied with the VCR or you can pick it up at any electronics supply store. The RF output can also feed another VCR, for tape dubbing, but direct video and audio connections (below) will achieve better results.

With regular VHS, the **composite video** connection will provide higher picture quality than an RF connection. With a Super VHS VCR, the **S-video** connection provides the highest picture quality, because it prevents the brightness and color portions of the video signal from interfering with one another. (For more details go back to "Television/Shopping for a DTV/Comb filtering.")

Elite videophiles with a D-VHS VCR should use the **IEEE 1394** connection, better known as FireWire, if possible. Unfortunately, finding something to connect it to can be a challenge. Many DTVs don't have IEEE 1394, and the only DTV tuner with an uncrippled FireWire connection has been removed from the market amid the baying and howling of Hollywood executives who do not want consumers to record videotapes in HDTV. In D-VHS, other high-quality output (but not input) connections include digital **HDMI** and analog **component video**.

The **stereo analog line audio** jacks are compatible with any receiver, as well as a second VCR, CD player, cassette recorder, or any other recording or playback device. You'll probably want to plug the stereo outputs into your receiver. Those two channels of analog audio are all it takes to carry the Dolby Surround signals that'll light up your receiver's Dolby Pro-Logic decoder.

The future of VHS

Though once the most widely used video format, VHS has become obsolete. It has been overtaken by DVD and Blu-ray in picture quality, hardware sales, and software sales/rentals. And its recording function has been usurped by hard-drive-based DVRs and to a lesser extent DVD recorders. The main reason for owning a VCR is to play an existing library. VHS VCRs are increasingly hard to find in stores, though you may find a few DVD-R/VHS combo recorders. JVC, which invented the VHS format, stopped making VCRs in 2009 except for DVD-VHS combos.

Shopping

Do the controls and menus make sense? Are you comfortable with the way tape moves—the transport's speed, rhythm, control structure? Are transport keys differentiated by size and shape at both deck and remote? Can your fingers find commonly used controls easily? Can the display (if any) be dimmed?

Useful Accessories

S ome accessories are afterthoughts. Others are bubble-packaged plastic marketing scams. But the right accessories, regardless of profit margin, are those that help your system achieve better performance or greater convenience. They enhance the value of what you already own and make it more pleasurable to use.

Remote controls

Nearly every time you add a component to your system, you add a remote control. After awhile they're collecting in a basket, or filling up drawers, or just scattered all over the place. At that point you wonder, wouldn't it be great if one remote could do it all? A good remote may not handle every single function, but will give you the basics—source switching, volume, mute, power. More sophisticated remotes really can do it all. Perhaps, though, a better goal would be to find a remote that'll just make life easier.

Universal remote control

A true **universal remote control** handles more than one brand of product. It may be packaged with an audio/video receiver, or some other product, or may be bought separately.

At its simplest, a "universal" remote may simply control other products of the same brand. Your TV may come with a remote that controls the same brand's disc players. That will work as long as you prefer to stay within that brand.

Thanks to HDMI, control automation has taken a further step forward with **CEC (Consumer Electronics Control**. It took effect with HDMI version 1.2a. Compatible DTVs, receivers, disc players, etc. communicate and coordinate commands as described above. Unfortunately manufacturers are muddying the waters by using their own proprietary names for this non-proprietary technology—such as Panasonic's Viera Link, Sony's Bravia Theatre Sync, Sharp's Aquos Link, Toshiba's Regza Link, etc. It's all CEC-based though interoperability among brands is not guaranteed. There are also some older pre-CEC control systems that enable components to jointly respond to commands.

If, like most home theater buffs, you end up with a multi-brand system, the most useful universal remote is the **programmable** kind.

Preprogrammed vs. learning remotes

There are two kinds of programmable remotes: preprogrammed and learning. Some remotes do both. Each kind acquires command codes in a different way. They may be bought as separate products or, if you're lucky, supplied with certain products such as receivers or TVs.

A **preprogrammed** remote has a built-in "library" of codes for different makes and models. Depending on the model, the database of codes may be updated to control products you add to your system in the future. A preprogrammed remote typically covers the most popular brands and models. If you think your TV brand is too obscure, you'd be surprised—there may be a preprogrammed remote that covers that brand. However, high-end audio products such as stereo preamps are not as well covered.

A **learning remote** acquires codes directly from the original remotes that came with your TV, DVD player, etc. When placed end-to-end with an original remote, it "learns" commands. The original remote emits the codes and the learning remote accepts them. A learning remote is inherently updatable—that's what makes it a learning remote. If you have a lot of legacy equipment, or plan to buy new equipment in the future, a learning remote will handle all those products.

Some remotes have both preprogrammed and learning capabilities. The preprogrammed side will speed setup, while the learning side will provide needed flexibility for special programming problems. A few learning remotes include some preprogrammed codes, but only for one brand (example: Marantz).

But most universal remotes are either preprogrammed or learning. Neither type is automatically superior. A preprogrammed remote requires you to enter the numeric codes that identify each of your components to the remote. You may be able to punch them in; or you may have to cycle through every code in the remote's internal database before finding one that activates the component.

A learning remote takes longer to set up. You're not telling the remote to recognize a make and model—rather, you're individually transferring each command that you intend to use. You'll quickly discover that transferring them all is rarely worth the effort and that front-panel controls have their uses. You may have to transfer some codes more than once to get them right.

Number of controlled components

A universal remote may control a small system or a large one. Get one that will handle all remote-controlled components in your system.

For a small bedroom system with only, say, a TV and cable box, a low-end remote that handles those devices will be adequate. But it will not cover a system that also includes an audio/video receiver and a rack stuffed with other components. Moderately priced universal remotes handle 8 to 10 devices. The pricier ones do well over a dozen.

Look and feel

Whatever kind of remote you buy, it should **feel** good in your hand. Rounder shapes typically have a better feel. However, the traditional flat shape lets you rest the remote on the arm of your chair, so you can punch the keys without having to hold the remote.

Button size involves a tricky tradeoff between simplicity and flexibility. When buttons are larger and fewer, the remote is easier to use—but can't do as much. When buttons become small and numerous, you can do more, but with greater difficulty. There are also some remotes that attempt to do a lot with a few buttons, relying on LCD screens. These can be wonderful or maddening.

Controls should be easy to tell apart by **size, shape, color, background color, layout,** or **labeling**. The most oft-used controls should be biggest. Volume controls should be large and easy to find. Certain traditional shapes work well for transport controls (play, pause, etc.). And there's no reason why all buttons have to be grey or black. Manufacturers who drown our fingertips in a sea of identical black-on-black buttons really need to invest a few more pennies in colored plastic.

Control layout also affects ease of use. Your fingers or thumb should easily find the volume up/down and mute buttons. Commonly used functions should not sit too close to controls that stop the show. Certain control settings belong under a lid, where they won't respond to an accidental touch.

Step-up remote features

Glow in the dark or **luminescent** keys make controls easier to see in a darkened home theater. They may be **backlit,** with a light source beneath the keys that glows briefly. The light source may be switched on manually (with a backlighting key, preferably side-mounted) or automatically (whenever you touch a button), giving you just enough time to

Logitech

Logitech's Harmony 700 ($149) makes it easy to set up one remote control to rule them all (your components, that is). Instead of an elaborate and trying button-pushing routine, this remote lets you enter information about your components into a web form. Then download the commands into the remote via USB and you're in business.

171

make up your mind and hit a key. **Phosphorescent** keys, like the hands of a watch, absorb room light and continue glowing after the room lights go out.

Macro keys—a top-of-the-line feature that has trickled down—store whole strings of commands. For example, you might set up one macro to turn on your receiver, select the BD input, power up the BD player, and extend its disc tray. Another macro might power down the whole system.

RF remotes are as handy as they are rare. They use radio-frequency signals in lieu of the infrared signals used by most remotes. RF signals can penetrate walls. Even when used within the same room, an RF remote is likely to have greater range than an infrared remote.

Touchscreen and PC-compatible remotes

Some higher-end remotes come with an **LCD touchscreen**. Unlike the liquid crystal display on your calculator or watch, a touchscreen responds to the touch of your fingertips. These so-called **soft keys** provide lots of extra flexibility. You can add or remove buttons, move them around, or rename them. You can even relabel them with icons. The remote becomes a willing servant that does whatever you want (if you're willing to work at it long enough).

The most versatile remotes are **PC compatible**. Some accept commands from the internet or a PC. The slickest ones, from Logitech, provide a web interface that focuses not on individual commands but on how you use your system. If you watch a Blu-ray disc by turning on your TV, receiver, and BD player, the interface will determine the necessary commands and the remote will download the commands. Then you can start a viewing session with one button. By substituting computer literacy for consumer electronics literacy, which is rarer, web-enabled remotes have become the wave of the future.

The newest idea is to use multi-purpose handheld devices as wi-fi enabled remote controls. Manufacturers are starting to provide free **apps** for the iPhone and iPod touch. Apps for BlackBerry and Android phones eventually may follow.

The best high-end remotes, once you've set them up and massaged the fine points of programming, make your whole system easier and more fun to use.

Upgrading and maintenance

You can always upgrade a learning remote. It's just a matter of placing it end-to-end with whatever remote comes with a new component. Transfer the codes and the original remote can go in a drawer with the others. (Keep them around in case you need to program a new learning remote.)

Upgrading a low to moderately priced preprogrammed remote may be possible in some cases. Some will accept new codes via CD-ROM or download. Others have to return to the factory to receive new **firmware** (software built into chips). Don't expect all cheap universal remotes to be upgradable.

If you need to change the batteries, don't worry about losing your programming. Most remotes come with a **battery backup** lasting from 90 seconds to 15 minutes. The remote will retain its codes long enough for you to replace the old batteries with new ones.

Power-line accessories

Products that mediate between house current and audio/video components take three forms: power strips, surge suppressors, and power-line conditioners. A power-line accessory may combine some or all of these functions.

Power strips

These simplest of power-line accessories do nothing more than convert one outlet to several, possibly with a fuse in between. Do not use a power strip in a home theater system. It will do nothing to protect your components, may inhibit their performance in some cases, and is often so poorly made as to constitute a fire hazard. (A tenant in my apartment house once caused a fire by using a power strip. I watched smoke and flames pour out of his windows. He left the building on a stretcher and was later evicted.) It's better to plug a source component or two into the back of your receiver than to use a power strip.

Surge suppressors

It's better still to use a *high-quality* surge suppressor. (Before we go any further, please note that the term "surge protector" is technically incorrect—a surge is not something to protect, but something to protect against.)

Good ones will protect your equipment from potentially damaging **surges**, **spikes**, or **transients** in the 120-volt power supply. That includes not only lightning strikes but more mundane power-line disturbances like an air conditioner's compressor switching itself on and off. Anything with a motor—a printer, copying machine, hair dryer, vacuum cleaner—will temporarily destabilize the power supply, possibly causing damage to delicate circuits in home theater components.

Most surge suppressors use **metal oxide varistors**. **MOVs** are semiconductor parts whose resistance drops as voltage rises. That enables them to soak up excess electricity and send it to ground. More than a hundred small surges per month can occur in an average household, so your power-line accessory's MOVs have their work cut out for them.

Some new products, including those from Tributaries, use **silicon avalanche diodes**. Unlike MOVs, which fail following repeated jolts, SADs send overvoltages to ground and trip a fuse. So instead of junking the whole product, you change a fuse.

When shopping for a surge suppressor, first look for **UL 1449** certification. This indicates that the product has been tested by Underwriters Laboratories for surge suppression (using 6000 volts of electricity!) and will not burst into flames if overheated. Because UL-certified products vary in performance—UL checks performance ranges, not specific benchmarks—you have more homework to do. There is also a **UL Standard 6500** certification which ensures that the product complies with all safety requirements of the **National Electrical Code**.

Next, check out the **joule** rating, a measure of how much electrical energy the product can absorb before it fails. You'll see a lot of three-figure numbers; for a home theater system, look instead for a four-figure number. And take it with a grain of salt. Some manufacturers make exaggerated claims concerning joule ratings. These ratings still don't tell the whole story.

Inflated joule ratings have led to an alternative rating known as **voltage let-through**, a measure of how much voltage the surge suppressor will conduct before it starts clamping. The less, the better. Even this specification doesn't tell the whole story because it fails to describe the shape of the surge's waveform. But the best choice for your system will have both a high joule rating and a low voltage let-through relative to other options.

Speed is also a factor. Look for some mention of **response time**. This is a measure of how quickly the product can react to protect your components. A response time of one nanosecond or less is good. A nanosecond is one-billionth of a second.

Power-line conditioners

Good power-line accessories—especially those calling themselves **power-line conditioners**—go further than just protecting your components. They filter out interference, isolate components from one another, and have status indicators for polarity and ground, enabling your components to perform at their best.

RFI or **radio-frequency interference** is an impediment to good performance. Basically, every piece of wire in your system—including interconnects, speaker cables, and power cords—can act as an antenna that picks up radio, TV, and other signals. Good power-line accessories filter RFI out of the power line before it reaches your components.

EMI is **electromagnetic interference**, caused when components radiate noise, either through the power line or through the air. The best power-line conditioners contain **isolation transformers** to prevent components from polluting other components through the conditioner itself.

Sequential turn-on is a useful feature when the power-line conditi-

Audio Power Industries

The Power Wedges from Audio Power Industries include isolation transformers to prevent components from polluting one another with interference. Shown is Model 1118 ($1449).

tioner is used in a large system. By slightly staggering the powering up of each component with a slight delay, it provides two benefits. First, it prevents a sudden massive power draw, with several components powering up at once, from tripping circuitbreakers. Second, it prevents switch-on noise from being heard through (or damaging) the system. Some components have circuits that make noises when they wake up. By placing your power amp after such components in the turn-on sequence, you can avoid that noise, sparing both speakers and ears.

Because electricians are human and can make mistakes, **indicators** are useful features. They may tell you whether the electrical outlet is correctly **polarized** (with the positive and negative elements in correct position) and **grounded** (to conduct electrical shocks safely into the earth).

Polarity is the reason why one blade of a two-prong plug is larger than the other. Inverted polarity isn't necessarily a safety issue but can cause performance problems. Ground—the third rounded prong in a three-prong plug—can be a safety issue.

Improper grounding can also cause a performance problem known as a **ground loop** when power and interconnect cords inadvertently set up a closed circuit between two unevenly grounded components. You may see the problem as a large noise bar rolling through the picture. Or you may hear it as a distracting hum.

High-end power-line conditioners (from companies like Audio Power Industries, Tributaries, and Monster Cable) can be quite costly. But that's because the best ones use hospital-grade circuitry to ensure that the performance of your costly components won't suffer. (Do athletes consume nothing but white bread and polluted water? Do world leaders dine on baloney sandwiches? 'Nuff said.)

Voltage stabilizers

A new and potentially helpful kind of power-line conditioner is the type variously known as a **voltage stabilizer**, **AC regenerator**, or **AC regulator**. These devices store power and ensure that connected devices receive a steady 120 volts. (That's for the U.S. The electrical systems of some nations, for example the U.K., differ.)

In the best of all possible worlds, that's exactly what your AC outlets would deliver. However, homes are situated varying distances from local power stations and those farthest away may receive a range of chronically inadequate voltages. When the power company is experiencing peak demand—routine in summer or winter—or damage to its plant, it may deliberately reduce voltage across the board, resulting in **brown-outs**. Closer to home, someone activating a power-sucking appliance in your home, or even next door, can cause a **power sag**.

These power-line fluctuations can cause performance problems. When a video display gets less juice than it needs, its power supply struggles to keep up, and the image may show flicker or video artifacts. Likewise a surround sound system experiencing a power drain may be subject to a sudden shrinkage or collapse of its soundfield.

To determine whether a voltage stabilizer would be helpful in your system, perform periodic spot-checks of your home's power supply using a **volt meter**. You can find an inexpensive one at RadioShack.

Uninterruptible power supplies

To protect against **sags** in the power supply, look into products that have an **uninterruptible power supply** (**UPS**) with **automatic voltage regulation** (**AVR**). They can prevent power from falling below or rising above a fixed level. These features are more critical for high-end computer users than for home theater systems. However, as audio/video products acquire more and more delicate computerized parts, UPS and AVR are starting to look like good ideas. Be warned, though, that a UPS capable of supplying enough current for a home theater system would cost serious money.

Correct use of power-line accessories

Don't plug a receiver or power amplifier into a surge suppressor or power-line conditioner unless it has a receptacle designed specifically for amps, which draw higher current than other components. Otherwise the dynamics of your system may suffer. Well-designed products will take this into account and provide one or more appropriately labeled outlets for receivers or amps. If you're a finicky audiophile, you might want to arrange loans from your local dealer, and listen to the sound of your amp before and after inserting prospective power-line conditioner purchases, to find what works best with your equipment.

Many surge suppressors come with insurance that will reimburse you if the surge suppressor fails and allows *properly connected* components to be damaged—note the proviso in italics. Every power cord, every video connection (antenna, satellite, cable), every phone connection (like the one on a satellite receiver), and every network connection (receivers,

177

Blu-ray players, streamers, servers) must go through the surge suppressor. Appropriate products will have receptacles for all these different kinds of connection. (Panamax products have a modular construction to allow subsequent add-ons.)

While a surge suppressor can handle the consequences of a nearby lightning strike, even a good one cannot defend against a direct hit. For maximum protection, make sure there's a **lightning arrestor** on your antenna, satellite, and cable feeds. Don't assume the installer will automatically provide it—some may neglect this important precaution.

Failing to make all the right connections is like walking out of your house and leaving the front door unlocked. It takes only one slip-up for an electrical intruder to get in and wreck the place.

Cables

A home theater system requires several kinds of cable including speaker cable, video interconnects, and audio interconnects (digital or analog).

Do you need high-end cable? As a rule of thumb, just be practical. Experiment and use what works best. It is true that speaker cables can sound very different from one another. (Electrical characteristics such as impedance and capacitance account for most of the differences. Unfortunately, companies marketing high-end audio cables almost never specify these things.) But don't rush into any rash, expensive purchases when you're setting up new speakers or amps and just getting to know them.

A lot of jargon surrounds the marketing of premium cables. You may see a lot of technical terms such as **dialectric** (insulation between conductors) and **skin effect** (current-flow pattern that skews frequency response). Manufacturers may boast about materials (oxygen-free copper is common, silver less so) and construction (the way the strands are wound or braided). While these things hold great fascination for advanced techies, most people are better off concentrating on the basics such as thickness, insulation, shielding, termination, and fireproofing.

If you're running cable behind walls, use the **fireproof** CL2 and CL3 designations, because cables can act as fuses, spreading fire. Consult a CEDIA technician and/or get a local electrician to pull cable.

Regardless of whatever kind of cable you're working with at the moment, try your best to minimize the length of all cable runs. Long cable runs degrade video and audio signals alike.

Speaker cable

The **thickness** of speaker cable is measured in **AWG**, which stands for **American wire gauge**. Speaker cable is most practical when it is generic and thick (at least 16AWG, and as much as 12AWG). As the gauge number drops, cable thickens. Thicker cables are more appropriate for the longest cable runs such as those to the surround speakers because they minimize signal loss.

One benefit of premium cable is **insulation**, to provide some measure of protection from abrasions, and **shielding**, to keep the signal from being polluted by **EMI**, or **electromagnetic interference**, from nearby components. If your system can receive a watchable picture from an indoor TV antenna, or a clean FM stereo signal, it may also be subject to **RFI**, or **radio frequency interference**, as cables can literally act as antennas, picking up stray noise. However, my some of my expert readers insist that such noise is well away from the audible spectrum.

If and when you are ready to invest in high-end speaker cables, the criteria listed above can be helpful in narrowing down your prospects to a tiny handful of options. I approached a major cable manufacturer to find one that was both 12-gauge and fireproof. Of the dozens of speaker cables offered by that manufacturer, only a few suited my requirements. When I added heavy insulation to my criteria, only two remained, and I ordered one of each (Monster M1.4S, biwired; and M1.2S).

Another benefit of premium cable is **termination**, or the sealing of bare wire tips inside various kinds of hardware. Copper, the material used to make most cable conductors, corrodes quickly when exposed to oxygen. Anything that can be done to seal it, even a few drops of solder, will extend the life of the cable. That in turn will protect your investment in premium cable and prevent you from having to restrip or replace it. Unterminated generic cable isn't necessarily a bad idea, especially when you're just starting your system, but it's not the best choice for a permanent installation that will go undisturbed for many years.

Cable termination may use hook-like **spade lugs**, plug-in **banana plugs**, or narrower **pin plugs**. Spades are the audiophile choice for binding-post terminals because they provide the greatest amount of **surface contact** (some people use the phrase **surface geometry**). However, the collared binding posts on most receivers cannot accept spades—in that situation banana plugs are a better option. As a reviewer who constantly swaps speakers in and out, I use bananas and they work well. If your speakers have cheap wire-clip terminals, pin plugs (along with soldered or bare tips) are the only options. Otherwise pins should be avoided because they provide the smallest amount of surface contact.

Video interconnects

Let's briefly review what past chapters have said about video interconnects. **HDMI** and **DVI** are high-quality digital HD-capable interfaces. **Component video** is analog but still HD. In the not-HD category are **S-video** and **composite video**.

The good news about video interconnects is that most of them come free with other things you buy. A Blu-ray player, if you're lucky, will come with an HDMI or component video cable and some audio cables, though not necessarily the ones you need.

More good news—three yellow-coded composite video interconnects of equal length can pass for a set of component video interconnects. One composite video interconnect can pass for a digital coaxial cable. Just make sure your substitute cables have a characteristic impedance of 75 Ohms. (Some yellow-coded cables are suspiciously slender.)

The bad news is that these free cables may be flimsy enough—with poorly fitting plugs, or broken strands—to require immediate upgrading.

HDMI interconnects (and hype)

As the HDMI interface steadily penetrates home theater systems, the quality of HDMI cables has become an issue. Here the marketing magic of the premium cable manufacturers is at its most intense. A lot of readers have asked how much to spend on HDMI cables. My best advice is to use whatever's supplied with your equipment. If you need more, check out oppodigital.com for sturdy, modestly priced HDMI cables. If your video display is 1080p, make sure it's a high-speed cable.

While some manufacturers have come up with elaborate hierarchies of HDMI bandwidth, there are only five official classifications. Here they are, straight from HDMI.org: "Standard HDMI Cable supports data rates up to 1080i/60. High Speed HDMI Cable supports data rates beyond 1080p, including Deep Color and all 3D formats of the new 1.4 specification. Standard HDMI Cable with Ethernet includes Ethernet connectivity. High Speed HDMI Cable with Ethernet includes Ethernet connectivity. Automotive HDMI Cable allows the connection of external HDMI-enabled devices to an in-vehicle HDMI device." That's all she wrote. Don't believe anything else you read on this subject.

HDMI cables carry great quantities of data. Like any kind of cable, they are subject to signal attenuation, and the longer they run, the more likely attenuation is to affect performance. Credible sources (outside the cable industry) say HDMI cable performance limitations can affect both picture and sound quality. Video problems may include noise and blocking. Audio problems may include jitter.

If you're running HDMI to a ceiling-mounted projector, a long run is unavoidable, and you may run into problems. Here are some solutions. One is to buy a thicker cable with higher quality of construction—this will expose you to marketing jargon and expense, but life is like that sometimes. Another solution is to use signal amplifiers, equalizers, or repeaters. Another is to convert HDMI to component video at one end and back at the other end. But the best solution, many installers say, is to use straight component video instead—unlike HDMI, it's analog, but like HDMI, it's HD-capable. And it stands up better to long runs. In any case, an experienced custom installer can help.

Monster Cable

The author's reference speaker cables are Monster 1.2 ($5/foot) and Monster 1.4s ($7.50/foot, shown) —the biwire version.

Digital audio interconnects (optical or coaxial)

Digital coaxial cables are so named because they are constructed with an outer sheath covering an inner core (a description that, coincidentally, also applies to the otherwise different RF cables that connect cable and other set-top boxes). Any yellow-coded composite video cable with characteristic impedance of 75 Ohms will do the job, in addition to the high-end cables made and sold specifically for digital use.

Digital optical cables send pulses of light through a filament. **Toslink optical cables**, which use a plastic filament, are the most common type. The plug snaps neatly into the jack. Optical connections get dodgy when the cables get kinked (treat them gently) or when jacks are contaminated with dust (keep optical jacks covered, and optical cables plastic-bagged, when not in use). There are also digital optical cables with glass filaments—they're occasionally used in high-end two-channel gear.

Audiophiles have long debated whether digital coaxial cables are better than digital optical ones. Manufacturers are concerned enough about the buzz to include both jacks on many products such as receivers and disc players. If there is a difference between optical and coaxial cables, I have never been able to hear it, and I've listened for it through some fairly good equipment. However, I have received a fair amount of passionate email from readers on this topic, and both camps make some good points. Coaxial cables have a higher bandwidth, and are certainly sturdier than optical cables. On the other hand, optical cables are immune from EMI and ground-loop hum. And the debate rages on.

Analog audio interconnects

In the analog sphere, interconnects are more likely to sound different from one another. If you're a high-end audio buff, and you've gone out of your way to buy a high-end preamp-processor or receiver with pure analog signal routing, high-quality analog interconnects are a necessary investment—both for oft-used source components, and especially for the amp/preamp interface. Also, if you're using the 5.1- to 7.1- channel analog line outputs of a Blu-ray, DVD-Video, DVD-Audio, or SACD player to feed a receiver or pre-pro, using high-quality cables makes a great deal of sense. For less frequently used components, premium analog interconnects are something you might consider fooling around with *after* you've matched your components, perfected your speaker placement, and paid off your mortgage.

For further reference

An excellent source of unbiased information on the controversial subject of cable is Stephen H. Lampen's *Audio/Video Cable Installer's Pocket Guide* (McGraw-Hill).

Racks, stands, & mounts

More than a mere afterthought, audio/video furniture should be considered another component in the system. It affects how your system looks—to you, family members, and guests—and even how it performs.

Audio/video component racks

Unless you want to buy a new rack every few years, make sure anything you buy can accommodate not only your current system, but anything you may add to it in the future.

Naturally, you don't want to bring anything ugly into your home. Is that "wooden" rack real hardwood or a veneer- or vinyl-covered particle board? There's nothing wrong with the latter. Lots of good entry-level racks are made of **medium-density fiberboard (MDF)**, an acoustically neutral material that's also used to make most loudspeaker enclosures. But it doesn't hurt to know what you're buying.

Some audiophiles prize steel-frame racks for their high rigidity and ability to control vibration. Shelves may be MDF, metal, or glass. I don't recommend glass, though it's common enough. Some glass racks have been known to spontaneously shatter, due to temperature changes, even when holding less than their specified weight.

Cable management features make it easier to add new components to your system. That's not to say that the easy way is the best way. Avoid **built-in power strips**. True, they make it easier to neatly arrange power cords, but a cheap power strip is no substitute for a high-quality surge suppressor—and may pose a fire hazard. A **built-in light** can be helpful, though, by illuminating the back-panel area where techies spend hours of our lives.

If you can't remove dust on, under, and around your system, it becomes a dirt magnet. **Casters** are one way to avoid that, though hardcore audiophiles prefer **spikes**.

Speaker stands

Choosing the right speaker stands is a question of performance as well as appearance. Flimsy speaker stands may subtly undermine your system's sound. Stands may have a **resonant frequency** that's audible when the bass guitar or kick drum hits a certain note.

In other words, they have a note of their own, and that note is audible when you rap on them with your knuckles. To prevent that note from affecting the musically significant midrange, the most committed audiophiles will do desperate things, filling **hollow** metal or MDF stands with sand or shot to muffle the stand's resonant frequency. (Sand and cat litter are safe choices. Lead shot may pose a health hazard.)

How the bottom of the stand meets the floor also affects sound quality. You have two choices, **spikes** or **feet**. Spikes couple the stand to the floor. They draw resonance out of the speaker cabinet, producing cleaner sound, and conduct bass through the floor, which can be quite exciting, both for you and your downstairs neighbors. To spare them the movie ballistics, use rubber feet instead.

Speaker mounts

Mounting a speaker is never sonically the best option. For good sound, stands or freestanding speakers are always best. Not all speakers sound good when placed close to the wall—in fact, most don't—but some are actually designed to be used that way. Unlike floorstanding speakers, small satellites, flat speakers, and bar speakers are often good candidates for mounting.

There are two main kinds of speaker mounts: keyhole and threaded insert. The **keyhole mount** is simplest. If your speaker has a keyhole on back, just sink a long nail into the wall, preferably into a stud, hang the speaker, and relax. Heavy speakers must be mounted into studs, not merely into drywall.

The best union of speaker and mount comes when speakers are equipped with one (or better yet, two) **threaded inserts**, preferably 1/4-20, the OmniMount standard. Some speaker models come with other sizes of threaded insert such as 4mm/8-32 or 5mm/8-32.

A speaker mount consists of a back plate that mounts to the wall, and a plate or other part that bolts to the speaker, usually separated by a hinge. The more metal parts, the better, although OmniMount does make a successful and popular hybrid with a plastic ball joint that swivels at a variety of angles.

Buying the speaker brand's mount is rarely a bad idea even if it costs a bit more. One popular speaker manufacturer designs all of its mounts to hold seven times the speaker's weight.

Be careful to compare the mount's **rated weight capacity** to the actual weight of each of your speakers, as specified in the speaker manuals. The safety-conscious consumer will buy a mount rated to hold several times each speaker's weight. Another benefit of buying a stronger mount is that the hinge will be stronger and therefore less likely to flop or wobble under the speaker's weight. You won't have to go crazy trying to aim the speaker in the right direction.

Consider what kind of **wall** you have. It may be drywall, double

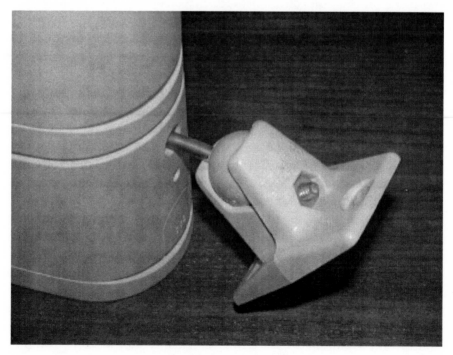

MF

The OmniMount 25RST-UMK ($59.95/pair list) is designed for small satellite speakers weighing up to eight pounds such as the Bose AM-5 shown here.

185

drywall, plaster, concrete, or brick. A typical suburban home has single or double drywall over wooden studs. One stud is located with a **stud finder** (an inexpensive device available at your local hardware store) and holes are bored to receive the mount's back plate.

Speaker mounts can be invasive, poking holes in walls and their underlying supports. If you're renting your house or apartment, speaker (and TV) mounting may not be a good idea. You might either violate your lease or leave a mess for the people who inhabit the space after you're gone.

TV mounts

TVs neatly mounted to the wall needn't be limited to pricey magazine spreads. Anyone who can knowledgably and precisely penetrate the wall of a house or apartment can mount a TV on wall or ceiling. If you can't do that, find an installer, dealer, or store that can. For larger sets and more problematic situations—for instance, you don't want to violate your lease—consider a TV stand instead.

As with any TV installation, **screen size** and **viewing distance** are interrelated. The viewer should be about three screen heights (or more) away from an HDTV.

A TV mount always has at least two parts: a **back plate** that mounts onto the wall, and a second plate, either a **front plate** to hold a flat TV or a **platform** to hold a direct-view TV. What happens in between affects both the mount's overall stability and its ability to move. Some TV mounts are **fixed**, while others **swivel** or **tilt**. Look for **cable management** features. Check **clearances** to ensure that the tilting TV doesn't hit the wall or cabinet doors.

Be absolutely certain that the mount's **rated weight capacity** is enough to safely hold both the TV's weight and, don't forget, that of the mount itself. Get your TV's weight from its instruction manual. The safest mounts are rated to hold several times the weight of your TV (and must live up to that rating!). With an adjustable mount, there is even greater need for the mount to support and manipulate the weight of the TV. Search online for product recalls and safety alerts, either on store sites or on the website of the U.S. Consumer Products Safety Commission: cpsc.gov/cpscpub/prerel/prerel.html.

As with a speaker mount, stop to consider what kind of **wall** you have, and how it may be penetrated, if at all. Mounting kits may allow for various types of wall, but TV mounting isn't a good idea for every wall. Drywall or plaster alone will not safely hold a large TV—you have to **penetrate a stud** behind the wall.

Some sets may be too heavy even for a **single-stud mount**. Then you'll need a **double-stud mount**, doubling the studs behind the wall. In new construction, it's a good idea to install double studs for future use. **Wooden studs** are stronger than metal ones. Metal studs are not advisable for large TVs. A **recessed mount** will need an additional wall plate for support and must maintain two inches of clearance all around for ventilation.

The more complicated the installation, the better off you'd be to leave it to an experienced and carefully vetted installer. If you need help, **CEDIA** (the **Custom Electronic Design and Installation Association**) knows someone who can do it right. For more information go to cedia.org.

TV stands

The right TV stand will safely support your set's weight—or more than its weight, to be certain—and will do so at the proper height for viewing. Consult your TV's manual to find out how much it weighs, and make sure anything you buy can support that weight, and then some.

Safety is an issue, especially in homes with children. A busy toddler loves to yank on cables and could easily pull a poorly situated TV down on its head. This results in tens of thousands of emergency-room visits every year. Your TV stand should be stable and solid, with wide legs or a solid base. The TV should sit low to the ground and near the back of the stand. Attach the set to the wall with safety straps or an L-bracket. Heavy items should go on shelves close to the floor.

Some TV stands have shelves made of glass. It's **tempered glass**, which disintegrates into pebbles (instead of angular shards) when broken. Despite the claims of manufacturers, some glass stands have been known to shatter under as little as half of their rated weight limits. Don't buy a glass stand.

For a bedroom TV, assuming it's too large to perch on a dresser, the simplest option is a **pedestal**, perhaps with a **swivel** to allow viewing from different parts of the room. But a TV used with source components needs a **rack** with space for components beneath the set.

The largest-scale option would be an **entertainment center** that holds the TV, components, and assorted media. These hulking things sometimes waste more space than they save, but if you want one, verify that it's a good fit for your TV's height, width, depth, and ventilation needs. Do not enclose a TV with side- or rear-mounted speakers in an entertainment center unless you don't plan to use them at all.

If the system includes a front center speaker, also consider where it will go—is there a shelf for it? Will the entertainment center leave room above or below the TV?

Connecting a Home Theater System

Quick guide

- Pick a hub
- Position components
- Speaker placement
- Subwoofer placement
- Connect video display
- Connect speakers
- Connect picture and sound sources
- Connect power
- Test system, source by source
- Run channel scan
- Arrange lighting
- Adjust picture settings
- Run auto setup and calibration
- Adjust surround levels
- Other surround settings
- Bass management settings
- Adjust subwoofer controls
- Program remote
- Acoustic tweaking
- Mechanical tweaking
- The upgrade itch

Installation guide

A good home theater system isn't built in a day. Setting one up is not something you do after getting home from work. Pick a lazy weekend when you've got time to do things right and attend to the fine points.

Some home theater components (for example, CRT projectors or high-end surround processors) are so sophisticated that they require a dealer's help to set up. Those buying high-end gear may need the help of a knowledgable installer to get the performance for which they've paid.

Pick a hub

Every system needs a central switcher. Therefore setting up a home theater system begins with a major strategic decision: Where will its heart be? Home theater systems tend to follow one of three patterns, connecting:

- through the television,
- through a receiver, or
- through a preamp-processor and power amplifier.

Connecting through the television set is the most convenient option. It's something you might do in a bedroom system, especially one whose dominant use is talking-heads telly. However, even practical home theater requires large enough quantities of picture *and* sound to create a true suspension of disbelief. That total immersion in the story is something the puny sound systems in TVs cannot provide. They can't compete with receiver-based sound systems.

The **audio/video receiver**, or **surround receiver**, is the logical system hub for a practical home theater. While it makes your system a little more complicated than a TV-centric system, a receiver-centric system brings several other significant benefits. First of all, sound quality will be vastly greater than anything a TV could provide. You have the opportunity to buy better speakers than those in your TV (and that isn't hard to do). A good receiver will provide enough power to make your five or more speakers sing. And it will throw in plenty of switching to handle your disc player, DVR/VCR, CD jukebox, PC/MP3 addiction, turntable, cassette, or satellite/cable/other boxes—as well as anything you might add in the future. Practical doesn't mean miserly!

More extreme addicts will gravitate to surround separates comprised of a **preamp-processor** and 5.1-channel (or more) **power amplifier**. Separating the amp's hot, power-sucking circuits from the preamp theoretically should let both do their jobs better, resulting in cleaner sound. However, such a sprawling system is not practical for everyone with its extra devices, cabling, and power consumption.

Position components

As you uncrate components, put the cartons into storage, and keep them at least for the warranty period of each product. You'll need them for any components that have to go out for servicing. If space is tight, flatten them and store the foam inserts in trash bags. Create a file for instruction manuals so that you can find them when needed.

Get all components in position before you start wiring them up. Don't just shove them anywhere. Stop to think about their ventilation needs. Consider which front-panel controls you'll be using most often. For example, if the disc player's play and eject buttons are located below the disc drawer, they'll be harder to reach if you put the player too low on the rack.

Place the video display (and seating) just beyond the point where scanning lines and dotted pixels (picture elements) are visible. If you're positioning a widescreen digital set, the minimum viewing distance is about 1.5+ times the screen diagonal; for an analog set, figure about 3+ times the screen diagonal. As long as you don't see the lines and dots that make up the picture, you've got it right.

Place other components where their interconnects can reach your receiver or preamp-processor. Be sure to allow adequate ventilation for all equipment—don't block ventilation holes on the top or sides. The best place for a receiver is atop the rack, where it'll have the most breathing room. If it must go lower, be sure to leave several inches for ventilation between the top of the receiver and the next shelf up. A gear closet will need a ventilation fan to keep components from overheating.

Speaker placement

Place speakers where the instruction manual recommends. (That may sound like a cop-out, but fact is, different speakers are designed to be used in different ways.)

You may wish to experiment with different placements of the oft-neglected front-center speaker. If it is on top of a rear-projection or direct-view TV, try angling it down toward the listening position by placing something under the back of the enclosure. If possible, better to put it below the screen, angling it upward. The aim, in either case, is to match tweeter outputs across all three front speakers, so the front soundstage maintains continuity when a sound pans from side to side.

Audiophiles prefer their main front left and right speakers a few feet from the wall because, for acoustic reasons, that's where they sound best. In some homes, audiophile-approved speaker placement may intrude into the room. But if you've invested in good speakers, don't give in to the compulsion to shove them up against the wall where they won't sound good.

Imaging—the system's ability to place objects accurately across the front soundstage—can be inhibited in home theater systems that place a video display between the front left and right speakers. The problem is **diffraction**, or sound waves bouncing off the TV. To minimize acoustic interference from the video display, place the front left/right speakers slightly farther out from the wall than the front center speaker, so that the three front speakers form an arc, rather than a straight line, and are equidistant from the listener.

Another highly effective way to adjust both imaging and tonal balance is to **toe in** the main speakers toward the listening position. This maximizes the proportion of sound heard on-axis. Some speakers are designed to work best when toed in; other are designed to sit parallel to the side walls. Toeing in generally produces more accurate imaging, along with stronger high frequencies. If you find the highs fatiguing, start reducing toe-in till you find the optimum tradeoff between accurate imaging and pleasing tonal balance. Another option is to put a wedge under the front of the speakers, or use lower spikes in front. That will not only reduce fatiguing highs, but will also minimize standing waves, because the sound will hit the back wall at a different angle.

If you listen often in stereo, using full-range front left/right speakers, positioning of those main speakers is critical. Bass will always be stronger when speakers are placed nearer the floor, walls, and corners. That's because such positioning will reinforce the tendency of bass waves of certain frequencies to bounce between parallel walls, a phenomenon called **standing waves**. Also, in different areas of the room, bass waves will either **reinforce** or **cancel** each other, something you can easily hear by putting on a low-frequency test tone and walking around. Systems heavily dependent on a single subwoofer (as opposed to two subs or full-range speakers) are especially subject to these effects. But don't assume you should always place speakers and subs to maximize bass. Too much bass can lead to a muddy, incoherent, and ultimately unsatisfying presentation. Instead, look for the right tradeoff between clarity and power.

Speakers that mate with flat-panel displays are increasingly popular. They're designed to mount on the wall, along with the display, so put them there. However, be sure to keep them all on the same level—above or (more likely) below the display. Do not put the left and right speakers alongside the screen unless you have only a stereo pair.

Surround speakers (as they are called in a 5.1-channel system) or side-surround speakers (as they are called in 6.1- and 7.1-channel systems) belong on the side walls, toward the rear of the room, while back-surrounds should be against the back wall. Their effects are supposed to be diffuse, so on-wall or in-wall mounting would be more acceptable. If you're into surround for music listening, back-wall placement is the rule. Experiment, if you like, but in a permanent 5.1-channel speaker installation, stick with side-wall placement. Later you'll experiment with speaker placement to adjust imaging, bass, etc.

A common flaw of many surround systems is that the rear speakers are aimed too directly at the listener. Bipole/dipole models avoid this by firing to the front and rear of the room, leaving the listener in the **null** area (the place with least sonic energy). To achieve the same effect with conventional (non-bipole/dipole) rear speakers, try aiming them at the ceiling. This works especially well if your system requires placement of rear speakers near the listener—say, on either side of a sofa. You can even do it with wall-mounted speakers. When the spray of sound hits the ceiling, it will be dispersed evenly and diffusely over the seating area. You'll need to set the receiver's surround levels a little higher to compensate for the reduction in direct-radiated sound.

If you are adding height and/or width speakers to your system, Audyssey recommends placing front left and right speakers 30 degrees off center, height speakers 45 degrees off center and another 45 degrees upward, width speakers at 60 degrees off center, and side-surrounds at 120 degrees off center. See diagram in "Surround speakers" chapter.

Subwoofer placement

Bass response from a subwoofer will be most powerful in corners because they emphasize midbass. However, corner placement is likely to give bass a booming one-note quality. It's better to keep your sub somewhere on the front wall, between the left and right speakers, or on the side walls, between the front and rear speakers. The sub in a compact sat/sub system should go as near the front-center speaker as possible because it has to deliver a higher range of frequencies. In that situation, you'll get a better blend if sub and center are close together.

Here's some tricks that'll help you find the ideal subwoofer placement: Put the subwoofer in your listening position and walk around the room till you find the spot where the bass sounds best. When you reverse the positions of yourself and the sub, the same acoustic conditions will apply, and the subwoofer will sound as it did when you first stepped into the magic spot. Another technique is to temporarily wire the main speakers out of phase (red to black, black to red) and adjust the sub's phase control till bass is weakest. Correct the speaker wiring to make it in-phase (red to red, black to black). This should put speakers and sub in phase with one another.

For a firm theoretical grounding in the science of acoustics, read *The Handbook of Acoustics* (4th Edition, McGraw-Hill) by F. Alton Everest. I also recommend a newer book, *Sound Reproduction: Loudspeakers and Rooms* by Floyd E. Toole (Focal Press).

Connect video display

If your system hub is a receiver or pre-pro, connect its monitor outputs to the video display. HDMI is the best choice in most systems. Newer receivers convert incoming video signals from any input and send them through the HDMI or component video outputs. In that case just make a single HDMI or component video connection to the video display. Older receivers may not convert video signals from one format (component, etc.) to another. These video streams are sometimes kept separate within cheaper or older receivers. In that case the display will need multiple connections, in order of quality: HDMI, DVI, component video, or (for pre-HD displays) S-video or composite video connections.

If you're connecting a front-projection display, be warned that not all interfaces work well with long cable runs. Best in this regard are 1394 and component video, both which can survive runs of up to 300 feet.

An RGBHV connection can travel up to 100 feet, an HDMI connection up to 75 feet, and a DVI connection up to 50 feet. In some cases signal convertors can extend these limits. Consult an installer.

Connect speakers

Premium speaker cable can be a good investment (if you don't overspend). But before you lay out big money for premium cable, start your system with generic wire ($15 for 50 feet). Using cheap cable first will give you a chance to perfect your cable runs, experiment with different configurations (such as biamping and biwiring), and avoid subsequent, expensive mistakes. Don't worry about pricey premium cable till you've broken in all the other components. Generic cable is, among other benefits, both inexpensive and fairly neutral. Do premium speaker cables make a difference? I believe they do, but you've got to graduate from high school before you go to college, so to speak.

With standard binding posts, use 16- to 12-gauge cable (thicker cable for longer runs) to connect your speakers. Note that as the gauge (AWG) number goes down, thickness increases. Measure each cable run from amp to floor to speaker and allow a few feet of slack. Cables for the three front speakers should match one another within a fraction of an inch. The same goes for the two (or more) surround speakers. Strip off a half-inch of insulation using a $10 wire stripper from the hardware store, an indispensable home theater accessory. It resembles a pair of pliers with different-sized notches for different cable thicknesses.

Spade lugs (Y-shaped hooks that fit around binding-post terminals) are reputed to provide the highest-quality connections because they have the largest area of surface contact. However, in my experience, flexing **banana plugs** sound just as good and, if they're snug, provide a more secure connection—spade lugs that come loose are potentially harmful to both people and equipment. Plugs are also favored by some speaker designers and reviewers because they make it easy to swap speakers and cables. Banana-plug adapters, sometimes called **double banana plugs**, can accept bare wire. If you want to use premium cable with sealed tips to feed speakers or a receiver with wire-clip terminals, get cables terminated with **pins** at the appropriate end. This provides a very small area of surface contact, so I don't recommend it—bare wire is better in this instance.

Speaker connections should always follow the standard color coding found on speaker terminals at both ends. Red always goes to red and black always goes to black. The red terminals are the **positive** (+) or **hot** connection, and the black terminals are the **negative** (-) or **ground** connection. More simply put, current goes in one side and comes out the other. This helpful color coding is also included on better speaker cables. If you're using generic cord, with two leads grafted together, you can figure out which end goes to which end by looking for a white or red stripe, or a ridge, on one side.

If you mismatch a speaker connection, the affected speaker will run **out of phase**. In other words, the drivers will move in when they should be moving out, and vice versa. This can result in a bass-cancellation effect audible as hollow, disembodied, unsatisfying bass response. If your first impression of your system is that something seems askew, disorienting, weird—then check your speaker connections immediately.

Biamplification allows a speaker with more than one set of binding posts to receive more than one channel of amplification. All other things being equal (such as the quality and quantity of amplification), biamplifying your system will definitely make it play louder and cleaner. However, biamping can cause problems when attempted with mismatched amps. Either the amps must be of the same design, or one must have a gain (volume) control so you can match them with a 1kHz test tone and a voltmeter. There are easier ways to make your system more dynamic—by upgrading the amplifier, or moving the speakers. Biamping should be thought of as a last resort for huge rooms that need more than one muscle amp to fill them up.

Biwiring duplicates runs of speaker cable to feed loudspeakers that have an extra set of binding posts, without additional amplification. This gives each driver a separate path to the amp, so the power-hungry woofer can no longer collapse the soundstage by drawing power away from the tweeter. The effects of biwiring are more subtle than the effects of biamping—in a budget system, don't bother. Biwiring is easier with premium cable—just tell the dealer that's what you want. But also keep in mind that biwiring and biamping double the amount of cable needed and therefore double the cost. Experiment with generic cable before investing in the fancy stuff.

Your receiver may have an **Ohm switch** on the back panel. It's designed to limit power output to prevent overheating. If your speakers have a low sensitivity rating (see "Surround speakers/Interpreting speaker specs") it might be wise to use the 8 Ohm setting to protect the receiver—actually, it might be an even better idea to arrive at a more appropriate pairing of speakers and receiver. Otherwise use the 4 or 6

Ohm setting to maximize power output. If your receiver is running not just warm but extremely hot, use the 8 Ohm setting, especially if the receiver shuts down more than once.

For the subwoofer, use either **line level** (low level) or **speaker level** (high level) interconnects. In other words, you may connect it using the same low-level interconnects you use to link other components to the system, or you may use speaker cables, which carry more current. The line-level (LFE or sub-out) connection from receiver to sub is usually best because it allows speaker-level connections (which handle everything above the sub crossover) to go directly from receiver to speakers. The speaker-level connection is occasionally a better choice for the smallest satellite & subwoofer sets because they are designed to function with a fairly high crossover from about 120 to 180 Hertz using the speaker-level crossovers in their subs.

Most subs come with both a low-pass filter and a high-pass filter. The **low-pass filter** is variable, and adjustable via the sub's crossover control. It ensures that the sub reproduces only sounds that fall below the crossover frequency. If the crossover control is set too high, you'll start hearing voices booming out of the sub. A receiver or preamp-processor contains a low-pass filter of its own which duplicates the one in the sub. If you were discerning enough to buy a sub that can switch out its own low-pass filter in favor of the receiver's, take advantage of that bypass mode. Otherwise your system's low-bass frequencies may suffer from having to pass through two filters, a process known as **cascading**. This is discussed in greater detail in the upcoming section on how to "Adjust subwoofer controls."

The **high-pass** filter is usually fixed, and used with a speaker-level connection. The receiver sends a full-range signal to the sub, where the high-pass filter strips out the lows, and passes the rest of the signal to the satellites. In rare cases, the high-pass filter may be used with a line-level connection. Either way, this may result in more dynamic sound, with less duplication of bass response—but at the expense of signal purity, because the signal has to travel farther, and through a filter.

If your receiver has more than five channels, you'll have to configure your receiver for use of the extra channels. This will be covered in a later section.

Connect picture and sound sources

It's time to link your system hub to signal sources such as Blu-ray,

DVD, DVR, VCR, cable, etc. Here are a few important rules for inter-connecting the components in your home theater system:

- Connect input to output, output to input.
- Use the same video connection throughout the signal chain for each component when using an older receiver.
- Connect video and audio to the same-named input.

First and foremost, observe the golden rule for any form of component interconnection: in-to-out, out-to-in. Do not plug an input into another input or an output into another output.

Next, note that some receivers require the video signal to travel over the same path from source component to receiver to TV. For instance, let's say your TV has HDMI input, but not component video. In that situation, you would not be able to connect the DVD player to the receiver using component video, because the receiver may not convert component video into HDMI for the TV. You'd have to use HDMI all the way through the video signal chain. Fortunately, an increasing number of receivers and preamp-processors provide **video upconversion** capability to translate one video mode to another, usually routing all incoming signals through the HDMI or component video outputs, enabling a single connection to the TV. That's progress!

Finally, group a component's video and audio connections under the same input name. If you connect your DVD player's component video output to "DVD," don't connect its audio outputs to "DVR." Most receivers or preamp-processors let you reassign inputs. Digital audio connections, like their analog counterparts, will be assigned to specific inputs. Whether a digital connection can override an analog connection made to the same input will depend on the design of your receiver or pre-pro. It may be necessary to go into the receiver's setup menu and assign an HDMI, coaxial digital, optical digital, or analog connection to each input.

Now, here's an overview of ways to make video connections:

- HDMI
- DVI-HDCP
- IEEE 1394-DTCP (DTV Link)

If possible, use these digital video interfaces—they afford the highest-quality connections. The best of the best is **HDMI 1.4a** and up, which handles all current video and surround standards. Unfortunately, older receivers don't have HDMI, so you'll have to connect set-top boxes and other video sources with these outputs directly to your DTV and/or video recorder. The **1394** interface is suitable for recording and networking; the HDMI and **DVI** interfaces are not.

If your system hub is a receiver, as opposed to a DTV, connecting video devices directly to the display will make switching more complicated. Audio connections still have to go through the receiver. So you end up switching video and audio separately. However, there is a viable workaround option—just program the multiple commands as a macro function into a programmable remote control.

In pre-HDMI components, the highest-quality video comes from analog connections that do the most to separate the video signal into its primary elements. It's best to use the highest-quality video connection supported by your video display. The best analog video connections, most of them found on receivers, are in descending order:

- component video
- S-video
- composite video
- RF coaxial

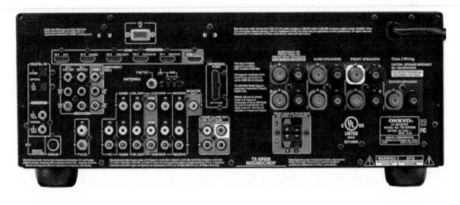

Onkyo

Back panel of the Onkyo TX-SR608 receiver. At top left is a row of HDMI jacks. You'll want to use them for as many source components as possible. Component and composite video jacks are also provided for older sources, but of the two, only component video is high-def-capable.

199

Component video is the highest-quality analog video connection found on a receiver, and the only HD-capable analog connection. It may be labeled in any of several ways including Y Pb Pr, Y Cb Cr, or YUV. Component video cable is often color-coded red, green, and blue to help avoid confusion. In any event, component video splits the video signal into three parts, including a brightness signal and two color-related signals. Three composite video cables with the correct characteristic impedance (75 Ohms) may substitute for a set of component video cables. When HDMI and DVI are not available, component video is the best way to connect an HD-capable video source to an HDTV.

S-video uses a multi-pin cable whose delicate pins separate video into brightness and color signals. By preventing brightness and color from affecting one another, an S-video connection can prevent visible problems such as cross-color interference and dot crawl. S-video is the highest-quality way to connect S-VHS VCRs and older equipment without HDMI or component video—but it is not HD-capable.

Composite video combines all the video components into one cable, color-coded yellow. It's useful mainly for connecting a VHS VCR or other legacy source, and is not HD-capable.

RF (radio frequency) coaxial cable carries multiple channels of picture and sound. It's used mainly to connect an antenna or cable feed to your system, or to feed an ancient TV set (via channels 3 or 4) that has only an antenna input and no direct video (S-video, composite, etc.) connection.

As for audio connections, HDMI is the best choice, digital coaxial/optical second best, and analog in last place—partly because most receivers convert all analog inputs to digital. Using direct digital connections helps avoid that unnecessary analog-to-digital stage.

Finally, here's a rundown on how to connect each source component in your system.

Blu-ray and DVD players: Use the highest-quality video connection supported by your video display, digital (HDMI, DVI, and 1394) or analog (component, S-video, or composite video). Tweakers seeking extra credit may wish to do some experimenting to determine which has best video processing: the player, the receiver, or the display. If you're using a 1080p display, start by having the player output 1080p. You can then set the receiver to process or simply pass through the signal. Then you can experiment with settings in the display. If the receiver is really mangling the signal and there's no quick fix, you might want to connect the player directly to the display for video switching, though this will result in awkward dual switching chores for video and audio.

If you're using a DVD player that upconverts standard-def material, another experiment worth trying is to feed the video display with either the player's progressive-scan output or its interlaced output. Here the aim is to see which works better—the player's video processor, or the one built into the display. Be warned that some progressive-scan players are shipped without the progressive output enabled—look for a button on the front or back panels or in the setup menu.

On the audio side, some (though not all) newer Blu-ray players use HDMI to output a bitstream to feed receivers with on-board decoding of the new lossless and other surround codecs. In that case, use HDMI. Dolby TrueHD, Dolby Digital Plus, DTS-HD High Resolution Audio, and DTS Encore are supported by HDMI 1.1 and 1.2 if your receiver has one of those inputs. For DTS-HD Master Audio your receiver will need HDMI 1.3. If your receiver does not have onboard decoding for the new codecs, but does accept high-res PCM, you can use the player's decoder that way with no loss in quality. If your receiver has neither on-board TrueHD, et al decoding or high-res PCM input, use the multichannel analog interface if it's available on the player.

To obtain old-school Dolby Digital and DTS surround signals from either a Blu-ray or DVD, HDMI suffices. For pre-HDMI receivers, connect one of the player's digital outputs to one of the receiver's digital inputs. For an older DVD-Audio and/or SACD player that has 5.1-channel analog line outputs and no HDMI, connect the 5.1-channel outs to the receiver's 5.1-channel ins. This analog connection may degrade the signal slightly, so a digital connection, when available, is preferable.

D-VHS VCR/HD-DVR: These digital video recorders require a direct link from your broadcast/cable/satellite set-top box using the IEEE 1394-DTCP interface—unless they contain ATSC (broadcast) or QAM (cable) tuners. Then they require an RF input. CableCARD-equipped gear requires an RF-in plus a CableCARD from your local operator. Use the highest-quality video output.

Analog VCR/SD-DVR: If you record only from an antenna, satellite, or unscrambled cable connection, you're free to link the video recorder to the receiver or TV as a playback-only device (S-video for an S-VHS VCR, composite video for regular VHS). But if you want to record from multiple sources, look for a VCR loop that includes both inputs and outputs. The antenna or unscrambled cable source should go to the video recorder's RF input (one of those big screw-threaded jacks with the tiny hole in the middle). You'll use the VCR's TV/VCR switch to select between tape playback or antenna/cable pass-through. See the section on cable TV (below) for more on the intricacies of the cable/VCR interface.

Video streaming device: These require a broadband internet connection. For output, use HDMI or the best quality alternative.

HDTV set-top box (broadcast, satellite, or cable): These devices receive RF inputs. They have both digital and analog video and audio outputs and which combination you use is up to you. An HDMI, DVI, or 1394 connection from source to DTV is the highest-quality video connection. If you prefer to switch the video output through a receiver, you can do so via HDMI or component video. The following sections refer to analog antenna, cable, and satellite connections.

Antenna: Use an RF splitter to divide the signal between the RF inputs of the DTV/set-top box and DVR/VCR. That will enable you to tape one channel while watching another, or to use single-tuner picture-in-picture to watch two channels at once. On a more practical level, it will also allow you to watch the news using the TV's built-in speaker(s) and without having to turn on the receiver. Don't buy an eight-way splitter if you're just splitting the signal two ways—each extra splitter acts as a signal-polluting broadcast antenna. There are two kinds of antenna connection: 75 Ohms, which mates with the sort of round coaxial cable your cable operator uses, and 300 Ohms, which mates with flat twin-lead antenna cable. The 75-Ohm type is better shielded and less susceptible to signal loss. A simple **balun** adapter, supplied with most VCRs, converts 75 Ohms to 300 and vice versa.

Analog cable box: Again, if you want to use the TV separately, or tape one channel while watching another, use a splitter to divide an unscrambled cable feed between TV and DVR/VCR. If your cable system uses signal scrambling on all channels, connect the cable box (or **convertor/descrambler**, as the cable guy calls it) between the cable feed and the recorder's RF input. That'll give you the freedom to tape programming without activating the rest of the system; or you can just watch cable TV through the recorder. Placing the recorder between the cable box and TV brings another benefit: it allows the recorder's MTS tuner to extract broadcast stereo audio (and embedded Dolby Surround) from the cable signal. If you want to watch one channel while taping another, things may get a lot more complicated. Most cable operators will make it easy for you—at slightly greater cost—by renting you a cable box with dual RF outputs. Otherwise you'll have to use an RF splitter and a (probably non-remote-controlled) A/B switch to select between the cable signal and some other viable video delivery source (antenna or satellite). Unless your favorite can't-miss programs are on at the same time, or you have a job that requires you to tape national crises on all major networks, this really isn't worth the extra complexity (and signal-smearing circuitry) that it inflicts upon your system.

Satellite: Connect the satellite feed—and the antenna feed, if applicable—to satellite and antenna inputs. Then, to get the signal out to your system, connect the highest-quality video output to a matching input on your receiver or TV—most modern receivers have one labeled "Sat." If your satellite receiver supports Dolby Digital surround sound, connect the Dolby Digital output to a digital input on the receiver. The phone jack must also be connected so that the satellite operator may communicate with your system for billing purposes during off-peak hours (usually late at night).

DVD-Audio/SACD: These high-resolution audio formats are best connected through HDMI 1.1 and up (DVD-A), HDMI 1.2 and up (SACD), or their 5.1-channel analog line outputs. Another alternative is to have the disc player output high-res PCM which can function with most HDMI-equipped receivers now on the market. Certain products from Denon, Meridian, and Pioneer have proprietary digital interfaces that work as well as HDMI. But ordinary optical and coaxial digital outputs, sadly, do not support full resolution. If you can't use HDMI, use the best-quality analog interconnects you can afford. The lack of bass-management features in older DVD-Audio products poses an additional hurdle—the format's weighty bass response might overwhelm some 5.1-channel speaker systems. Either upgrade to a new model or add one of the outboard bass-management boxes (but only if excessive bass is an acoustic problem in your space).

CD: Unless you're certain that your receiver is designed to process analog signals in the analog domain, it's safe to assume that it will convert all analog inputs to digital. So no matter how good your CD player's digital-to-analog conversion is, you'll never get the benefit of it—might as well bypass the receiver's A/D stage by using a direct digital connection, coaxial or optical. The D/A converter you hear will be the receiver's. On the other hand, if you chose your CD player partly on the basis of its D/A performance, and you're sure the receiver won't convert it to digital, feel free to experiment to see which connection (digital or analog) sounds better. The result is not easily predictable. Trust your ears. By the way, using a digital connection will also make it easier to record from CD with a digital recorder (CD-R or MiniDisc).

iPods/MP3 players: A few receivers accept iPod input via USB. For the remainder, use an iPod docking device. Some are optional and designed to work with specific brands of receiver. Others have standard a/v outputs and can work with any receiver. To plug in any brand of portable audio player, buy an adapter that has two RCA-type plugs at one end (for your receiver) and a 1/8-inch stereo mini-plug at the other end (for your player). Plug it into an unused stereo analog line input in back of the receiver. You can also listen to MP3s through a home theater system using a BD/DVD/CD player that's compatible with MP3s burned onto CD-Rs.

Digital audio server: Use the HDMI or optical/coaxial digital outputs for the highest-quality connection. Your DAS may also require a network connection for internet use or home networking.

Cassette: Does anyone still make cassettes? OK, then use the analog tape loop jacks to connect the recorder's outputs to the receiver's inputs and the recorder's inputs to the receiver's outputs. Otherwise, you're free to treat the cassette deck as a playback-only source and it'll plug easily into any pair of stereo analog line audio jacks.

Phono: Not just any old analog input will serve a turntable. Why? Because phono cartridges produce a tiny voltage that's not nearly as great as the output of, say, a CD player. You'll need a dedicated phono input, something no longer found on every receiver, to amplify that delicate signal and correctly process it using an industry-standard equalization curve. If your receiver does have a phono input, unless specified otherwise, it will be compatible with **moving-magnet** phono cartridges, the more common and less costly kind found on affordable turntables (and favored by the hiphop community for better bass and durability under the stress of "scratching"). A few receivers have phono inputs switchable between moving magnet and **moving-coil** cartridges, which produce more accurate highs, and generate even tinier voltages. If you have a large library of LPs and/or 45s, and your receiver has no phono input, don't lose heart—just add a separate **phono preamp** to your system, plugging it into any available stereo analog audio input.

AM/FM: Connect the supplied antennas per the instruction manual. Most receivers use threaded 75 Ohm RF-type terminals. In areas with weak FM reception, you may want to (further) split your TV antenna feed and use it for radio. Do TV antennas pick up FM radio? They sure do—all FM stations lie between channels 6 and 7 in the VHF-TV band. However, be warned that some TV antennas (and cable systems) use **FM traps** to strip out the FM signals, because they may interfere with VHF-TV channels.

XM or Sirius satellite radio: Some receivers have inputs for next-generation satellite radio, either XM, Sirius, or both. You'll need to buy an external antenna for about 20 bucks. And you'll need to activate an account. Then you can select satellite radio as a source and listen to hundreds of channels.

Network connection: Connect broadband to the receiver's ethernet input to access streaming video, internet radio, media from a PC, etc.

For more information on connecting picture and sound sources, see the "Connection glossary."

Connect power

Don't connect power cords until you've done all other connections. Avoid running power cords in parallel with video or audio cables—the power cords will radiate electromagnetic noise into the interconnects.

It may be tempting to connect all power cords to the same power strip or surge suppressor. However, if the power accessory is not of high quality, it can degrade performance. Strongly consider adding a good **power-line conditioner** to your system. A top product will protect components from surges, isolate picture and sound sources from one another's signal-polluting influence, and tell you what's going on with the power line (whether it's properly polarized and grounded).

If you're getting by without a power-line conditioner, at least refrain from plugging a receiver or power amp into anything but the wall, so it can draw all the current it needs. Use your receiver's **unswitched power outlets** to connect components that need a constant supply of juice (DVR, satellite receiver, etc.). The **switched power outlets** can be used for the other components. Note that most components have power on/standby modes rather than a simple on/off arrangement—don't plug these into a switched outlet. They need power to retain settings.

Try to avoid connecting components to a power line that's shared by an air conditioner. As the AC's compressor switches itself on and off, surges or dips may occur in the power line, which could inhibit the performance of your components and possibly even damage them.

Test system, source by source

Now the fun begins—sort of. Turn on the receiver, set it to the

AM/FM tuner, and dial in your favorite station. Even static will do. It's better to start with the tuner than with, say, the disc player because if there's a problem, you'll know it's not the source component. Then work your way through all the other components. Make sure a/v sources produce a picture and fill all appropriate channels with sound.

As you do this, keep in mind what you connected to what. Some of your video sources may have been connected through the receiver, others directly to the display. You may have to switch video and audio separately. Later you'll automate that double round of switching by programming your remote control with macro commands. But don't worry about that right now.

Don't be upset if you don't get everything right on the first try (I rarely do). Deal with arising problems by checking connections. If the physical connections are in place, start going through the menus of the receiver or pre-pro as well as other menu-driven components (such as the DVD player or DVR). To get everything to work you may need to use the menus to activate, reassign, or rename certain inputs or outputs.

If you get no sound at all from the main speakers, don't panic. A receiver that accommodates a second pair of speakers (over and above the usual 5.1-channel surround array) will have an A/B speaker switch, and you may need to activate it to get any sound at all. In addition, some 7.1-channel receivers allow the back-surround channels to do double duty for height or width speakers, second-zone use, or biamping of the front channels. You may have to switch them from one use to the other. If that's not the problem, start looking for loose or missed connections.

For more snares and solutions, see "Connecting a Home Theater System/Problem solving."

Run channel scan

Now that your source components are running—but before you start dealing with picture and sound quality issues—this would be a good time to run the **channel scan** for your TV and any other source components with tuners (DVRs, set-top boxes). Definitely do this if you're using an antenna or a CableCARD-equipped TV. If you are a cable subscriber, but receive only the cheapest broadcast-channels-only service, you might want to try this with the cable connected directly to the set—if your TV has a QAM tuner, it can handle unencrypted channels. If you're using a cable or satellite box, you don't need to do this.

First, make sure your antenna or cable feed is connected.

You may be asked to specify whether the signal source is an antenna or cable feed, and whether to skip unused or unavailable channels. Start the scan and the device will then detect any analog and digital channels available. There may be a manual option but auto is usually easier.

Be warned that some stations fail to transmit accurate channel-associated data. This may cause certain channels to display inaccurate information (or no picture at all). There's not much you can do about it but complain to the station.

When the channel scan is complete, you should be able to grab the remote and flip through all available channels. That can be enormous fun. Maybe this would be a good time to rest from the rigors of system setup, kick back, and relax for awhile!

Arrange lighting

Depending on the kind of video display in your system, lighting will have to be adjusted.

A flat-panel or direct-view TV can more easily survive in a well-lit room than a projection set. However, it will look its best in a room with low light. Too much light forces you to raise contrast and brightness to levels that shorten the display's lifespan. Too little light forces the eye to focus on both a bright picture and a dark background. That is likely to lead to eyestrain and headaches.

The light should not be allowed to reflect off the screen. So the best place for lighting is behind the set. And placing a shaded low-wattage bulb (25-40 watt incandescent or 7-10 watt fluorescent) behind the screen will bias the optic nerve, making movie-length sessions more comfortable. Yes, you can actually tweak your own eyes for better performance! For extra credit, raid a well-stocked lighting store for bulbs with a color temperature of 6500 Kelvins (the packaging may specify it). The CinemaQuest Ideal-Lume is designed especially for this purpose.

For wall-mounted sets, a few well-aimed ceiling spots may help. Again, make sure other room lighting does not reflect off the screen

Projection sets are another story entirely. Because they don't produce as bright a picture as direct-view tube TVs, they work better in a darkened room. CRT projectors in particular may require blackout curtains or blinds, especially for daytime viewing. The Eclipse Thermawave is well made and does a fine job. For best results, choose a dark color.

If you've invested big bucks for a front-projection display for use *in a dedicated home theater room*, paint the walls black or dark grey (avoiding glossy finishes) to give the projector and screen the best possible chance to work their magic without light reflecting from the walls and ceiling. However, be warned that dark walls make the room impossible to use for any other purpose—they suck up so much light as to make the room dim and dismal. In a multi-purpose room, consider darkening just the wall with the screen.

Get a handle on room lighting before adjust the video display.

Adjust picture settings

Here's a quick guide on how to adjust your video display for best performance—and longest life.

First, realize that your home is not a showroom. Once you get your set home, it won't be competing with a wall of other screens. So there's no reason to stick with factory settings designed to help a set compete under bright showroom lights. In fact, factory settings typically overdrive projector lamps, LCD backlights, and tubes. With tubes, the results are especially bad, distorting shapes onscreen, and imposing needless wear and tear. Even if you're not using a tube display, factory settings will make your display look awful.

Start by setting the display for its **movie** or **cinema mode**. This will immediately get most settings close to ideal.

For manual adjustment, begin by backing off the **contrast control** (or **white level**). It adjusts the balance of relative black between the brightest and darkest areas of the picture. Adjusting it either makes light areas lighter, or moves everything toward a uniform shade of grey. You're likely to find the correct setting below the factory preset (which may be cranked all the way up). The best setting might even be below the manufacturer's center position, though TVs vary. In direct-view sets and CRT projectors, excessive white level causes the picture to lose sharpness as the picture tubes are overdriven. When that occurs, the result is **blooming**, as the dotted phosphors grow fat and spill light into adjacent areas. Other signs of excessive contrast include bending of vertical white lines and brownish discoloration of bright areas that linger on screen for more than a few seconds.

Brightness (or **black level**) adjusts the amount of total black in all areas of the picture, light or dark. Too little brightness (or too much black level) brings a picture that is too dark, and starved of shadow detail, as objects in darker areas disappear into black. Too much brightness washes out the picture. A correctly adjusted picture has a deep black—as close to the absence of light output as you can get. True black is more achievable in plasmas and tube-based displays than with LCD and microdisplay technologies.

The **color** control adjusts the overall **saturation** level (intensity) of each of video's three primary colors (red, green, and blue). Again, the factory setting may be too high. Look for a bright red object (the red color bar in a test pattern is ideal but anything will do). Reduce the color level until the red object's edges cease to bleed into adjoining areas. This works well for analog TVs, which have trouble reproducing red.

The **hue** or **tint** control (the technical term is **color phase**) changes the balance between red and green. Its primary use is to adjust flesh-tones, among the most difficult colors to reproduce accurately. However, fleshtones vary, and to complicate matters, some sets come with fleshtone adjustments that make Caucasian people look right but skew all other colors. So also give some attention to the green playing field in a sporting event (preferably natural turf).

Finally, the **sharpness** (or **detail**) control affects the definition of edges and fine patterns within objects. Most people set it too high in the mistaken belief that they're seeing more detail; what's actually happening is that they're seeing more noise. The correct setting is likely to be from 0-30 percent for Blu-ray/DVD or satellite reception, since these sources produce a very clean picture, and slightly lower for analog video sources. Some sets can memorize different picture settings for each input—a very useful feature that saves time for fussy videophiles.

The second best way to fine-tune your picture settings is with a Blu-ray test disc such as *Digital Video Essentials: HD Basics* (Joe Kane Productions) or the *Spears & Munsil HD Benchmark*. These are useful tools, both for setting up a new video display or tuning up an old one.

But the videophile who wants everything just right will bring in a technician licensed by the **Imaging Science Foundation**. ISF techs calibrate your set using professional test equipment, burrow into control menus not accessible to the user, and make fine-tuning adjustments beyond the skill of an average person. The service may be included in the cost of a high-end video display or you may bring in the technician on your own. The cost is a few hundred dollars—but if you've spent thousands for a high-end video display, it's well worth it. Even moderately priced TV sets can benefit from ISF calibration. It can make the difference between a mediocre picture and a great one. For more information about ISF, visit imagingscience.com.

Run auto setup and calibration

Most receivers can set themselves up automatically and implement room correction to fix acoustic problems. Unless you're an advanced user, you'll want to run the program. If you don't like it, you can always turn it off later.

Before you run auto setup, make sure that your speakers are in the right places. Place a tripod in the prime listening position, attach to it the small microphone that came with the receiver, turn on the receiver, and plug in the mic. If the receiver doesn't automatically start the auto setup program, start it from the control menu.

The program may ask you some preliminary questions about how many speakers you have and where they are placed. In a seven-channel receiver, the sixth and seventh channels may be configured for back-surround, height, or width speakers. If your speaker setup is basic 5.1, you can also use the extra channels to biamplify the front left and right speakers. Or they can be adapted to multi-zone use. Receivers with more than seven channels allow more of the above options. You may also need to set the sub's volume knob before auto setup begins. Turn it one-third to halfway up. You can always change it later.

As the program runs, your speakers will emit test tones. Be sure not to stand between the mic and the speakers or make any noise—that will throw off the measurements. The test tones may be loud, so either use hearing protection or, better yet, leave the room.

Most auto setups now take measurements from more than one listening position to provide better response outside the traditional sweet spot. You don't have to run the maximum number, but it's a good idea to cover all the main listening positions. If the audience rarely spills off the sofa, take three measurements on the sofa and call it a day.

When the measurements are complete, the program may take a few minutes to calculate the right settings for your system. Then it will probably offer you a look at the settings. Make sure the speaker and sub distances look right. If the speakers are detected as "large" or "full range," and you'd rather have the sub handle frequencies below 80Hz—as recommended by THX—then set the speakers as "small" and select the proper crossover.

If you later move your speakers, even by small amounts, even just a single speaker, run auto setup again. Without correct calibration, a surround sound system can't deliver the enveloping feeling that suspends disbelief and pulls you into the story. Instead, by calling attention to itself, it just competes with the movie. We master technology so that art can take precedence over technology.

Adjust surround levels

As an alternative to auto setup, some surround buffs prefer to fine-tune channel levels manually, using a sound pressure level meter. You'll be using it to calibrate the sound levels of each speaker to ensure a uniform surround soundfield. It should be possible to do all this from the remote while seated in the center listening position.

- Turn on meter to BATT setting to check battery.
- Set range control to 80.
- Set weighting curve to C.
- Set response to SLOW.
- Put receiver into surround mode.
- Activate receiver's test tone.
- Adjust each channel 5dB above 80 within 1/2 to 1dB.

Turn on the meter. Before you do anything else, check its battery strength, to make sure it will provide accurate readings. Then set its **range** control to 80 decibels. Set its **weighting curve** to C. (C-weighting covers all frequencies. The alternative setting, A-weighting, rolls off steeply in the midrange and bass.) Finally, set the meter's **response** to SLOW to stop the needle from jumping around too much.

Put the receiver (or pre-pro) into one of the surround modes and activate the test signal. It's **pink noise**, spanning a wide range of frequencies, and sounds like a continuous storm of static. The signal may jump from channel to channel, or it may stop on each channel—some receivers will allow you to choose the mode. You're aiming for a volume level of 85dB, or 5dB above 0 when the meter is set at 80. (You could also read it as 5dB *below* 0 when the meter is set at 90.) Try to get each channel within 1dB of 85 (some processors allow increments as small as 1/2dB). This method works for THX and other gear where the master volume control does not affect the level of the test tones. However, on some receivers and pre-pros, the master volume control does affect the volume level of the test tones. If that's the case, set the front left channel at 0dB, and turn up the master volume until the meter reads 85dB. Use that as the benchmark to balance all the other channels.

Perfect settings may be unattainable, but if you must err, do so on the plus side for the front-center channel and on the minus side for the rear speakers. You want the front-center speaker to deliver intelligible dialogue; you don't want the surrounds to call attention to themselves.

Repeat this process each time you add new speakers to the system, or even when you just move a single speaker.

Other surround settings

Even in receivers without auto setup, setting **delay times** has become easier. Getting delays right used to require the user to calculate and choose the correct number of milliseconds based on positioning of speakers and listener. Nowadays, rather than make you struggle to understand what the settings mean, a good receiver will simply ask you how far you're sitting from each speaker, and make the appropriate choices based on your responses. Key in the distances and you're done—no muss, no fuss, no mathematics.

Setting the surround delay time adjusts the soundfield between the front and rear channels to ensure that ostensibly simultaneous sounds from each speaker arrive at your ears at about the same time. In Dolby Pro-Logic mode, where sound tends to leak between channels, there's a second reason for the delay—it minimizes **Haas effect**, the perception that whenever sound arrives from two sources, you perceive it as coming from the closer source. Delay ensures that, even if you're sitting physically closer to the surround speakers, sounds that are supposed to come from the front will sound that way.

Bass management settings

Bass management settings direct low bass frequencies—the hardest ones to reproduce—to the speakers best suited to handle them. Auto setup will handle this. If you'd prefer to second-guess auto setup, tell the receiver whether each of your speakers is **large** (with full-range woofers) or **small** (satellite speakers that need a subwoofer for bass reinforcement). The size of the speakers is really less of an issue than their frequency response; a "large" speaker is one that reproduces frequencies below the crossover frequency, usually 80Hz. You'll be asked whether the system has a subwoofer (a yes/no question).

One common setup would be "large" front left and right speakers and "small" center and surround speakers. If the main speakers provide enough bass to eliminate the need for a sub, the subwoofer channel is shut down altogether ("off") and bass frequencies are routed to the main speakers. (The THX people, among others, encourage you not to run "large" speakers with a sub. Some THX gear will prevent you from doing this, though in most cases it's more a suggestion than a requirement.)

One feature showing up in a handful of products is the ability to mix bass from the front left and right channels into the subwoofer output. This is a way of second-guessing the mixing engineer. Try it, if available, and see if you prefer it.

If your receiver has power to spare—which is rarely the case—and you're doing without a sub, you may want to run all of your speakers full-range. Simply set the sub to "off" and everything else to "large." You'll get the soundtrack's full bass response all around. (Unfortunately, many soundtrack mixing engineers elect to omit low bass frequencies from the surround channels under the assumption that most systems won't have full-range speakers in the rear.) At the opposite extreme, systems that use small satellites all around cannot function without a subwoofer, and with these systems the receiver must send *all* bass to the sub. Set speakers to "small," subwoofer "on."

With powered towers, the main speakers' side-mounted active drivers effectively function as subwoofers. The built-in subs may receive either direct input from the receiver's subwoofer output, if they have line-level inputs (speakers "small," sub "on"), or may derive the subwoofer signal from the main left/right speaker connections (speakers "large," sub "off"). Check the speaker manual.

Adjust subwoofer controls

Nearly all subwoofers come with a few standard controls. The **volume** control, of course, adjusts the sub's internal amplifier relative to the rest of the sound (as do the receiver's setup menu and volume knob). You can try using the sound meter to measure the sub's output (be sure it's set to C-weighting to include bass frequencies). But you'll probably get wildly different readings in different parts of the room. If you're like most people, you'll start out by setting the sub volume too high because, after all, you paid for the thing and you want to hear it. But over time, you'll discover that subwoofers work best when they disappear into the soundfield, providing only subtle reinforcement most of the time, and playing loudly only at peak moments. Feel free to experiment; I predict you'll end up with a lower setting than you started with.

The **crossover** control determines the frequency at which the sub takes over bass-producing duties from the satellite speakers. Read the specs for your satellite speakers to find out where their bass response begins rolling off. Set the crossover too low and you'll end up with a **notch** (or dip) in midbass response. Set it too high and the area of frequency duplication between sats and sub will be audible as a **peak**, which can sound muddy and oppressive. Room acoustic conditions may affect the crossover setting.

Most systems will have subwoofer crossover controls both in the surround processor (receiver or pre-pro) and in the sub itself. If the signal passes through both crossovers, an undesirable effect called **cascading** will result. In the best of all possible worlds the subwoofer would have a bypass switch (or a separate input that bypasses the crossover) and you'd set the crossover in the surround processor's menus. However, if the sub doesn't have a bypass, and the signal must pass through both crossovers, open one of them up all the way by setting it as high as possible. That will minimize the cascading effect.

Which crossover is better? That may spend on the **slope** of the satellite speakers—in other words, how steeply they roll off bass frequencies. Most crossover controls don't adjust slope. If your sub and sats come from the same manufacturer, the sub crossover's slope may be more appropriate than the one in your surround processor.

The **phase control** helps smooth out differences in bass response that occur when the subwoofer is a different distance from the listener than the main speakers—as it often will be, for either performance or aesthetic reasons. You want all sounds to reach your ears at the same time. Adjusting the phase control determines the direction in which the sub's drivers move, which in turn affects how bass is propagated from the sub. The phase control may be continuously variable, or may have just two settings. Experimentation will determine which phase setting produces better bass.

Program remote

Depending on what kind of remote you've chosen, programming it may take some doggedness and dedication. A preprogrammed remote requires just a few steps to identify your components to the remote. But getting a learning remote to do everything you want will take longer.

The inner workings of remotes are variable enough to make it impossible to give step-by-step instructions here. Yes, you'll have to read the instruction manual, even if you're a man, and program the remote with the manual by your side. But here are a few tips to get you started.

Remote capacities vary, both in the number of components handled, and the number of functions for each component. Don't drive yourself crazy trying to get your learning remote to do everything—there are times when it's easier to use front-panel controls or the original remote. Instead, prioritize. Concentrate on getting the universal remote to handle the most regularly used functions.

Once you've got the basics under your belt, you can start taking advantage of macros, which shoot multiple commands at the press of one or two buttons. That can be useful because there are sequences of commands that come up over and over. For instance, when you're getting set to watch a DVD, you'll power up your DTV, receiver, and player, select the DVD input on the receiver, and eject the drawer to receive the disc. Such repeated command sequences are prime candidates for automating via macro commands.

Macro commands are also a great way to organize multiple rounds of switching in a complex system. If some video sources connect to the receiver, and others directly to the display, you shouldn't have to remember that every time you switch from one source to the other—make the remote do it. If video and audio require dual switching for some sources, let the remote do that too. A little time invested in programming macros today will save you a lot of embarrassment in front of guests tomorrow.

Acoustic tweaking

Tweaks—defined as small things you can do to produce minor improvements in system performance—can be **acoustic** or **mechanical**. Of the two, acoustic tweaks are by far the most effective.

The room is not just an empty vessel to be filled with people, technology, movies, and music—it's another component in the system. Reproduced sound is a combination of **direct** and **reflected** sound. In a concert hall, the majority of the sound reaching your ears is reflected. In a home theater, it's better to cut down some of that reflected sound, and try to get more direct sound.

There's an entire industry devoted to acoustic tweaks based on highly advanced (and costly) technology. However, you can improve the sound of your space without hiring an acoustician by doing a few simple things.

The most damaging reflections are those from the floor and the

215

side walls. Reflections from these areas make the sound abrasive and incoherent (not exactly a winning combination). So controlling reflections from these areas will help prevent room acoustics from spoiling the surround sound presentation, which is so critical to the suspension of disbelief that gives the home theater experience its emotional power.

First: to stop sound from reflecting off the floor, get a rug. Cover all or most of the floor with it. And make sure there's a carpet pad beneath it so that the end result is extremely thick. (I keep an old shag rug under my oriental rug; it's like walking on a cloud, and is very acoustically absorptive.) The most famous practitioner of the rug strategy is President Jimmy Carter, who tells a story about the time he invited Vladimir Horowitz to play at the White House. The pianist listened to the acoustics of the East Room and refused to perform until the president had gone upstairs to fetch a carpet. Carter, the perfect host, obligingly moved the oriental rug around the room until Horowitz was satisfied with the acoustics. Floor damping will benefit Republicans and Democrats alike.

Next, to stop sound from reflecting off the side walls, cover up the side walls just in front of the main speakers to the extent possible. An acoustic treatment doesn't have to look like an acoustic treatment— thick drapes or shelves stuffed with books work beautifully. Otherwise, consider tapestries, wall-hanging rugs, egg carton (whose variable surface diffuses side-wall reflections), or a thick sheet of foam rubber. Even an object just one or two feet square can make a big difference in taming side-wall reflections.

If the seating is right up against the wall—as sofas often tend to be in midsized or smaller rooms—treat the wall behind the listening position with something that diffuses sound. (Egg carton is effective, albeit unsightly. It's OK cover it with a curtain or tapestry.) Otherwise the wall will reflect too much bass, and muddy everything else.

How do you know when to stop deadening the room? While a 5.1-channel home theater system requires more damping than a 2-channel music-only system, it is possible to go too far. A good way to get a quick fix on room acoustics is the **slap echo test**. Just clap your hands and listen to the sound as it bounces off the walls and decays. If it decays quickly, with very little reverberation after the initial slap, the room is in good shape. If you hear several distinct sounds a fraction of a second after the original, you may have work to do, and that can be fairly obvious—your own voice will sound echoey too. Don't over-do it. You'll want to have a little room reverb left to make music sound right. In some over-damped home theater installations the reverb is a weird *boink* that makes music sound lifeless and unnatural.

Finally, pay attention to how your furniture affects the dispersion of sound. Avoid high seat backs that prevent sound from the side- and back-surround speakers from reaching your ears. Your system won't have a chance to sound good if the chair you're sitting in undermines it.

Mechanical tweaking

Killing mechanical resonance is an audiophile obsession that has inspired a whole genre of accessories that do one of two things. Either they dampen mechanical energy, by transforming it into tiny amounts of heat, or they simply move it elsewhere. These are controversial tweaks. Not everyone agrees that they work, or how they work.

Energy dampers often look like hockey pucks—you'll even see them serving as the feet of receivers and other products. Placed under components, they absorb vibration that otherwise would pollute delicate circuits.

Energy couplers, in contrast, are usually steel cones or spikes that soak up energy on one side and transmit it through their sharp tips. Depending on whom you ask, their primary benefit is either to drain mechanical energy from components, or to prevent floor-borne vibration from affecting components. They may be used for just about any component (a disc player, an amplifier) but are most commonly found in loudspeakers, which often have threaded inserts on bottom to accept spikes. You might enjoy the physical sensation of low bass being transmitted through the floor and up through your seat. But the real reason to spike your speakers is remove extraneous music-polluting energy from their enclosures. Spikes do an especially good job of quelling the riot of vibration in a subwoofer enclosure. A spiked sub produces cleaner bass pitches.

The upgrade itch

Today an average lifetime encompasses several generations of changing technology. Therefore it's not impractical to consider home theater system upgrades every few years. High-definition video technology is taking off like a rocket. And new lossless technologies are starting to transform surround sound for the better. A home theater system in step with the times is open to change.

There are intelligent upgrades and there are stupid upgrades. In the latter category are inferior HD and sub-HD flat panels, muscle amps far more powerful than smallish speakers and rooms require, a set of speakers that resembles Stonehenge and hogs the room, and cables that cost more than major components (though adequate cable is a necessity). At the other extreme, the smallest satellite speakers and cheapest receivers are subject to the laws of physics and economics, and to hope otherwise is wishful thinking. As Lancelot Braithwaite observes: "Bigger is not necessarily better. Smaller is not necessarily better." And while we're at it, even flatter is not necessarily better.

Upgrades that are worth it improve your system's performance in a manner that average eyes and ears can appreciate. That includes the quality of video and audio but there's more to it than that. For home theater to be a viable lifelong passion, it must fit into a practical life—or several lives if you have a spouse, family, or roommates.

So don't just think of ways to make your home theater system bigger or smaller. And don't feel obligated to chase the latest fad format (unless you're likely to collect a substantial library in that format). Rather, think of ways to make your system more elegant, stealthier, easier for other household members to use, more ... practical. Replacing a big butt-ugly analog television set with a high-def flat panel or projector, or replacing the huge speakers you bought in your foolish youth with smaller ones that sound better—these are the smart ways to a upgrade home theater system. The job, if you love home theater, is never really over.

Connection glossary

Following, in numerical and alphabetical order, is a list of connections and associated terms used in home theater systems.

1/8-inch mini-jack/plug: Though rarely seen in home theater systems, these space-saving connectors are common in PCs and portables, including MP3 players. You'll need a mini-to-RCA adapter to connect a PC's soundcard or portable device's headphone jack to your system.

1/4-inch headphone jack/plug: The standard in home-audio headphone connections (originally known as the *phone* jack/plug and used by early 20th century Bell System operators).

5C: Former name of *DTCP* (see separate entry).

5-way binding posts: Screw-down speaker terminals that accept spade lugs, banana plugs, double banana plugs, pin plugs, or bare wire. *Collared* binding posts do not accept spade lugs.

5.1-channel to 7.1-channel inputs/outputs: Six to eight analog RCA-type jacks provided to connect SACD, DVD-Audio, and some Blu-ray and DVD players. Also used to connect a surround preamp-processor to a multi-channel power amp.

12-volt trigger: Used to send control signals between components, usually to coordinate startup.

75 Ohm antenna input: The type used on most TVs. Converts to 300 Ohms with an adapter.

300 Ohm antenna input: The type used on older TVs. Converts to 75 Ohms with an adapter.

A/B speaker terminals: Some receivers allow connection of a second stereo pair.

Analog: A signal carried as a continuous waveform (as opposed to a *digital* signal, which is carried in pulses representing zeroes and ones).

ATSC: The standard-setting body that originated the DTV standard, the Advanced Television Systems Committee, also gave the standard its cryptic acronym of a name. ATSC is the digital TV standard in the U.S., Canada, and Mexico; *NTSC* was the analog TV standard.

Audio: Usually refers to a pair of RCA-type analog line-level audio connectors, but could include digital types.

Audio/video: Generally refers to a pair of RCA-type analog stereo line-level audio connectors (color-coded white and red), usually mated with a one or more video connections.

Audyssey: A company that licenses auto setup, room correction, low-volume listening modes, and width/height-enhanced listening modes.

Audyssey DSX: An Audyssey-licensed pair of listening modes. One adds width channels to a standard 5.1 array. The other adds height channels. They can be used separately or together. They also can be used along with back-surrounds to bring the total number of channels in a system to 11.1.

Audyssey Dynamic EQ: An Audyssey-licensed listening mode that aids low-volume listening by adjusting frequency response and surround levels.

Audyssey Dynamic Volume: An Audyssey-licensed listening mode that aids low-volume listening by reducing the ratio between soft and loud.

Audyssey 2EQ: An Audyssey-licensed auto setup and room correction program that takes measurements from up to three room positions.

Audyssey MultEQ: An Audyssey-licensed auto setup and room correction program that takes measurements from up to six room positions, or up to 32 for custom installers.

Audyssey MultEQ XT: An Audyssey-licensed auto setup and room correction program that takes measurements from up to eight room positions, or up to 32 for custom installers.

Audyssey MultEQ XT32: An Audyssey-licensed auto setup and room correction program with the greatest filter resolution of any Audyssey auto setup technology.

AWG: American Wire Gauge, a measure of the thickness of cable. In home theater it applies mainly to speaker cable. An average gauge is 16. The gauge number drops as thickness increases, so 14AWG is thicker than 16AWG, and 18AWG is thinner. Gauges higher than 18 or lower than 12 are not advisable.

Banana plugs: Inch-long plugs that easily connect speaker cables to

MF

Above are some jacks you're likely to see on the back panels of various components. From top and left: wire clips (PSB), 5-way binding posts (JBL), biwire binding posts (Platinum Audio), multi-use RCA-type inputs (Pinnacle), Toslink digital optical inputs (both uncapped and capped, McIntosh), XLR balanced analog line input (McIntosh), S-video jacks (Zenith), RF terminals (Zenith). In the lower righthand corner, from top, are IEEE 1394, HDMI, and DVI jacks.

speakers and amps equipped with binding posts. The flexing type provides the best fit—other kinds tend to fall out of the post. May be built into *double banana-plug adapters*, which are helpful for converting bare wire tips to bananas.

Binding posts: Speaker terminals that accept a variety of specially terminated cables. Opposite of *wire clips*. *Five-way* binding posts accept spade lugs; UL-approved *collared* binding posts do not.

Biamp/biwire speaker terminals: Allow connection of two sets of speaker cable. Biamplification adds extra amp channels to provide speakers with more power. Biwiring gives each speaker driver a direct path to the amplifier; its benefits are more subtle.

CEC: Consumer Electronics Control. It enables components connected via HDMI 1.2a (and up) to coordinate commands. Though it is a non-proprietary standard, CEC is often disguised with proprietary names such as Panasonic's Viera Link, Sony's Bravia Theatre Sync, Sharp's Aquos Link, Toshiba's Regza Link, and others.

Characteristic impedance: A laboratory measurement that measures a cable's resistance to current while assuming that the cable is of infinite length. That's impossible, of course, but the aim is to specify the characteristics of the cable itself and not what it's connected to.

Codec: Encode/decode format, for audio/video or audio only.

Collared binding posts: Speaker terminals that accept banana plugs, pins, and bare wire but not spade lugs. *Five-way* binding posts do accept spade lugs.

Coaxial (vs. optical): Usually refers to a digital audio connector using an RCA plug or jack. Also frequently used to describe the 75 Ohm RF-type cable used to deliver cable TV. More generally, refers to any cable with an outer sheath surrounding an inner conductor.

Component video: A direct analog video connector that separates three different parts of the video signal, a brightness signal and two color-difference signals, usually with RCA-type connectors, color-coded red, green, and blue. Better than S-video, composite video, or RF-modulated

Monster, Tributaries, MF

Here are some plugs and other termination hardware often found on the tips of inter-connect and speaker cables. From top and left: HDMI (Monster), IEEE 1394 (Tributaries), DVI (Monster), component video (Tributaries), S-video (Tributaries), RF coaxial (slip-on type at top, screw-on type at bottom), the multi-use RCA plug (XLO), Toslink digital optical, XLR (Esoteric), spade-lug speaker cable (MIT), banana-plug speaker cable (Straight Wire), pin-plug speaker cable (Monster).

video signals; not as good as HDMI or DVI. May be labeled *Y Pb Pr*, *Y Cb Cr*, *Y B-Y R-Y*, or *YUV*. Used to connect DTVs to set-top boxes, DVD players, and other signal sources.

Composite video: An analog video connection using an RCA-type connector, usually color-coded yellow. Not HD-capable.

DBA-25: Multi-pin connector that carries 5.1-channel analog line-level signals. Invented by THX. Rarely used.

Digital: A signal carried in pulses representing zeroes and ones (as opposed to an *analog* signal, which is carried in a continuous waveform).

Dolby Digital: Encode/decode format widely used for surround sound in theaters and home theaters. Usually feeds a 5.1-speaker array, though Dolby Digital EX decoders can derive a sixth channel to feed one or two additional back-surround (rear center surround) speakers. Used in DVD, satellite TV, and DTV broadcasting.

Dolby Digital EX: An addendum to mainstream surround formats that derives a sixth channel from a 5.1-channel Dolby Digital signal. This extra back-surround channel typically feeds two speakers placed at the rear of the room. Developed by both Dolby Laboratories and THX, first licensed by THX as *THX Surround EX*, and now licensed by Dolby too.

Dolby Digital Plus: An improved version of Dolby Digital, a form of efficient lossy surround. Found in some Blu-ray disc releases.

Dolby Pro-Logic: Analog surround decoding format. A Pro-Logic decoder expands two channels of input to four channels of output that feed five speakers and a sub. There is no Dolby Pro-Logic signal or audio connection per se. Pro-Logic decodes Dolby Surround signals which may travel via the stereo soundtracks of MTS broadcasts, videotapes, or the two-channel alternate soundtracks of some DVDs. All Dolby Digital gear includes backward compatibility with Dolby Pro-Logic and Dolby Surround.

Dolby Pro Logic II: A new and improved version of Dolby Pro-Logic that is suitable for use with stereo source material with or without embedded Dolby Surround.

Dolby Pro Logic IIx: The 7.1-channel version of Dolby Pro Logic II, with back-surrounds.

Dolby Pro Logic IIz: An enhanced version of Dolby Pro Logic II and IIx, with height channels.

Dolby Surround: Analog surround format that is encoded in the stereo soundtracks of videotape and some videodisc releases as well as MTS broadcasts. Back in the 1980s, there were Dolby Surround receivers and processors for home use that fed four speakers. However they were supplanted by Dolby Pro-Logic equipment, which uses the same signal but adds a front center speaker. All Dolby Digital gear includes backward compatibility with Dolby Pro-Logic and Dolby Surround.

Dolby TrueHD: A form of lossless surround, the highest-quality codec available from Dolby Labs. Found in Blu-ray disc releases.

Dolby Volume: Low-volume listening mode that evens out volume levels among content sources while maintaining listenable audio quality.

DTCP: Digital Transmission Copyright Protection, a copy-prevention scheme, supported by several TV makers, that prevents unauthorized copying through the IEEE 1394 digital interface. One of two copy-prevention schemes used in an increasing number of DTV displays, set-top boxes, and recorders. *DTV Link* is an IEEE 1394 interface with DTCP.

DTS (Digital Theater Systems): Alternative encode/decode format for surround sound in home theaters and movie theaters. It is lossy, which means it omits some data to save bits. The 5.1-channel version is used on some DVD-Video, DVD-Audio, and CD titles. Competitor of Dolby Digital. Also called *DTS Digital Surround.*

DTS 96/24: A high-resolution audio-only DVD format from DTS.

DTS Encore: The original 5.1-channel version of DTS surround, used as a designation on software.

DTS Express: Secondary audio format for DTS-HD encoded Blu-ray discs.

DTS-HD High Resolution Audio: A newer version of DTS surround. It is still lossy, like the old version, and therefore efficient, but produces higher-quality surround. Found in Blu-ray disc releases. Competitor of Dolby Digital Plus.

DTS-HD Master Audio: A form of lossless surround, the highest-quality codec available from DTS. Found in Blu-ray disc releases. Competitor of Dolby TrueHD.

DTS Neo:6: A DTS mode that rechannels stereo sources to up to 6.1 channels, usually served by a 7.1-channel system.

DTV Link: See *IEEE 1394*.

DVI-D: Digital Video Interface, created by the computer industry's Digital Display Working Group, to handle digital signals. The DVI signal is not uncompressed, contrary to popular opinion, though it does have huge amounts of peripheral data deliberately added to make it too unwieldy for digital recording or networking. DVI signals are no better or worse in quality than component video signals. Other versions include DVI-A, for analog signals, and DVI-I, for both digital and analog signals. An adapter can *physically* mate DVI to HDMI but the two connected devices will work together only if they share the same signal protocols, chiefly copyright protection.

DVI-HDCP: Digital Video Interface with High-Definition Copy Protection. The latter is heavily supported by Hollywood, and is one of two copy-prevention schemes used in an increasing number of DTVs and set-top boxes. Now superseded by *HDMI*, which is smaller and supports audio as well as video. DVI converts to HDMI easily with an adapter.

F-connector: Plug found at the tips of the RF cables used for cable TV and other video delivery sources.

FireWire: See *IEEE 1394*.

Fixed output: An audio output whose signal cannot be adjusted by the volume control. The opposite of *variable*.

Front-panel inputs: Jacks located on front of TV or other product, convenient for camcorder or videogame connections.

HDMI: High-Definition Multimedia Interface, a new combination of *DVI* plus multi-channel audio, stereo audio, integrated remote control, infrared repeater, and IP signals. Despite its all-embracing nature, the connector is actually smaller than the DVI connector, and promises to untangle much of the rat's nest of cables that plagues home theater systems. The original *HDMI 1.0*, released in 2002, supports video and stereo audio, not surround. *HDMI 1.1* adds Dolby Digital Plus, Dolby TrueHD, DTS-HD High Resolution Audio, and DVD-Audio. *HDMI 1.2* adds Super Audio CD. *HDMI 1.2a* adds CEC. *HDMI 1.3* came out in 2006 and supports DTS-HD Master Audio, higher bandwidth, greater color depth, a new mini-connector for camcorders, and automatic lip-sync to keep soundtracks coordinated with faces speaking onscreen. *HDMI 1.3a* and *1.3b* make small modifications relating to compliance testing but offer the consumer no functional improvements over HDMI 1.3. *HDMI 1.4* adds 3D, IP device sharing, data exchange, other features. *HDMI 1.4a* has the fullest set of 3D options for movies, broadcasts, and games.

IEEE 1394: Generic designation for a high-speed digital connector available in four- and six-pin versions. The four-pin version is more common in audio/video products, the six-pin version in computer products. Also known under such proprietary names as Apple's *FireWire* and Sony's *iLink*. Named for the Institute of Electrical and Electronics Engineers, which originated it, plus the document that contains its specifications. *DTV Link* is an IEEE 1394 interface with DTCP copy-prevention features.

iLink: See *IEEE 1394*.

Impedance: The load that a speaker presents to the amplifier. *Nominal impedance* is the specified number, though in practice impedance varies with the signal, and routinely drops below that number. *Characteristic impedance* relates to cables.

Line level: The standard range of signal voltages used by receivers, disc players, and many other home theater system components.

Monitor output: On surround receiver or preamp-processor, an

HDMI, component video, S-video, or composite video output feeding a video display.

MTS: Multi-channel Television Sound, the official name of the standard for stereo analog television broadcasting, invented by Zenith. Usually includes dbx noise reduction and always accompanies SAP (Second Audio Program). May survive in analog cable channels.

Multi-room: Audio system serving more than one room. Also called *multi-zone.*

Multi-room/multi-source: Audio system serving more than one room with more than one simultaneous audio or a/v feed.

Nominal impedance: Speaker specification that describes a speaker's resistance to current and therefore the load that it presents to an amp.

NTSC: The North American television standard, named for the standard-setting body that originated it, the National Television System Committee. Adopted in 1941 and updated for color in the early 1950s, NTSC was the analog TV standard in the U.S., Canada, Mexico, and Japan; the U.S. discontinued it in June 2009. Replacing it is *ATSC*, the digital TV standard in the U.S., Canada, and Mexico.

Optical (vs. coaxial): Usually refers to a digital audio connector using fiber-optic cable. Some believe digital *coaxial* connections to be superior.

PCM (pulse code modulation): A generic type of digital audio. Uses and resolution vary. For instance, a Blu-ray player may convert lossless surround to a high-res PCM signal for a receiver that lacks TrueHD decoding.

Phono input: On a receiver or preamp, an input that accepts the fragile low-level audio signal coming from a turntable and phono cartridge. Boosts and equalizes the signal for listening on a home audio system.

Pin plug: A slender form of speaker-cable termination that mates with wire-clip speaker terminals.

RCA (vs. XLR): Connectors using the quarter-inch-wide plug and jack invented by RCA Laboratories. Sometimes referred to as *unbalanced* (nothing wrong with that, just a technical term). RCA jacks are widely used in audio and a/v gear for analog audio, digital audio, composite video, and component video connections.

RF (radio frequency): In television, a 75 Ohm connection that carries multiple channels of video and audio, often used for satellite/cable connections.

RF-modulated digital input: In home theater, the rare special input required for a receiver or preamp-processor to accept Dolby Digital signals from a laserdisc player.

RGB+HV: A form of analog component video found in high-end, pro, and computer gear. Includes red, green, blue plus horizontal and vertical sync.

RS-232: A computer-type port used in custom installation to coordinate the operation of touchscreen interfaces, firmware upgrades, and other functions.

SAP: Second Audio Program, the ability to switch a TV broadcast from one soundtrack to another. Helpful for bilingual broadcasts among other uses.

Spade lugs: Y-shaped speaker connectors that mate only with binding posts (they will not fit wire clips).

S/PDIF: Sony/Philips Digital InterFace, the digital audio connection used in receivers and other digital audio/video products. May be *optical* or *coaxial* types.

Speaker level: Connection using speaker cables, usually referred to in connection with subwoofers.

Stereo: Two-channel signal. One mode of many on an audio/video receiver. The way most music has been recorded for the last four decades. An aesthetic that brings together low-end compact-system users and high-end audiophiles. In its original form (now archaic), the word *stereophonic* referred to anything that provided a three-dimensional sonic perspective.

Subwoofer: A speaker, usually with internal amplification, that produces only low bass.

Surround sound: In its most minimal (and standard) form, surround is a 5.1-channel format, five speakers plus a subwoofer. Height, width, or back-surround channels may raise the total to 7.1. With all possible options, the maximum under current technologies is 11.1 channels.

S-video: An analog video connection that splits the signal into brightness and color. May be labeled Y/C (Y is brightness, or luminance, while C is color, or chrominance). Better than composite video but not HD-capable. Recently eliminated from some receivers and other products.

THX certification: A certification program for speakers, receivers, and other home theater components that ensures they meet performance benchmarks.

THX Loudness Plus: A low-volume listening mode that reduces the ratio between soft and loud sounds.

THX Multimedia: THX certification for desktop use.

THX Select, Select2, Select2 Plus: THX certification for smaller rooms.

THX Surround EX: See *Dolby Digital EX*.

THX Ultra, Ultra2, Ultra2 Plus: THX certification for larger rooms.

Toslink: Official name of digital optical cable with plastic (as opposed to glass) filament. Toslink jacks are found on receivers, DVD/CD players, digital audio servers, and other source components.

Variable output: An audio output adjustable through the volume control. The opposite of *fixed*.

VGA: 15-pin computer monitor connector.

Video: When used without qualification, the term usually refers to composite video connections, but could also refer to other types.

Wire clips (spring-loaded or clamp-down): Speaker terminals designed to receive bare wire, usually 16-gauge or less, and not much else. Universally considered inferior to *binding posts*.

Y/C: A video connection that separates *Y* (luminance/brightness) from *C* (chrominance/color). See *S-video*.

XLR (vs. RCA): 3/4-inch-thick connector using a three-pin jack or plug. Carries line-level audio signals for long distances, either in professional installations or high-end home stereo systems. Rare in surround gear. Also referred to as *AES-EBU*, having been adopted by both the Audio Engineering Society and the European Broadcasting Union. Sometimes referred to as *balanced*.

Problem solving

You can find instruction manuals for many products on manufacturer websites, often in Adobe Acrobat format (the Acrobat reader is a free download at adobe.com). Here—loosely based on the "troubleshooting" guides often found in those manuals—are ways to solve some of the problems that arise in home theater systems. What kind of problems? Well, problems with...

TV sets

The first paragraph of advice given here may seem insultingly obvious, but according to tech support people, it's one of the most common problems they hear about. I've refrained from repeating it for other products, though it applies widely.

No power.
- Check power on/off button.
- Is it plugged in? Just asking!
- Check AC outlet or power strip slot by plugging in a lamp.

No picture or sound.
- Is input source on? Try another input on TV.
- Check selected input on receiver.
- Press TV/VCR button on VCR (it selects between antenna/tape).
- When using cable or satellite box through RF connection, switch from channel 3 to 4 or vice versa.
- Check video menu settings on TV, receiver, and source (especially DVD) to ensure that signal is getting through.
- TV requires separate antenna/cable/satellite connection when not used with receiver.

No picture.
- When routing video through receiver, use same video-connection type throughout.

No sound.
- Check audio menu settings on TV, receiver, and source (especially DVD) to ensure that signal is getting through.
- Component, S-video, and composite connections do not carry audio (only RF carries both video and audio).
- When viewing DTS DVD, turn on DTS output in DVD player menu.

Color is unnatural.
- Adjust color or tint controls.
- Adjust "color temperature" or "color balance" to 6500 Kelvins (or "low"/"natural" setting).
- Consider having your TV calibrated by an ISF technician.

Scanning lines too visible.
- Adjust viewing distance to five times height of screen (analog 4:3 TVs only).
- Lower detail control (you'll also get rid of lots of video noise that way).

No response to remote.
- Aim remote directly at player.
- Aim remote at a less extreme angle.
- Inspect batteries for correct polarity (+/-).
- Replace dead batteries.

Big-screen DTVs

Big-screen displays have distinct problems of their own. Here are some.

Colors bleed along edges of objects.
- Some projection TVs require "convergence" of tubes or other imaging devices—consult manual or technician.

Dim picture.
- Projectors need ambient-light control. Darken room, with blackout curtains if needed.
- Try a screen with higher "gain" (reflects more light back to viewer).
- If problem is most severe at sides, a lower-gain screen will have a wider "sweet spot."
- Projector lamps quickly lose brightness early in lifespan. Again, dark room is best solution.
- Reduce size of front-projected picture.

Painfully bright picture.
- Most TVs are factory-set way too bright and contrasty. Turn these settings down.
- DLP and LCD projectors produce brighter picture than other kinds. Again: turn settings down.
- DLP and LCD projectors can benefit from a lower-gain screen.

Oddly shaped picture.
- Adjust aspect ratio (screen proportions) to suit programming: 16:9, 4:3, letterboxed, anamorphic, etc. This adjustment is found in DVD players as well as DTV sets.

Make those lines and dots go away!
- Cut back on detail control.
- All tube and some projection TVs have visible scan lines. Increase viewing distance or decrease projected area.
- Some fixed-pixel displays have visible pixel grid. Reduce detail.
- Consider adding a video scaler to your projector.

Diagonal edges are jagged, motions distorted, backgrounds shift, other weird stuff on-

233

screen.
- Cheap line doublers (built into HD-RPTVs) cause video artifacts. A progressive-scan DVD player might help.
- Up- or downconversion between video formats can cause unpredictable effects.

Blotches mar large areas of the screen.
- This problem is inherent in many gas plasma displays. It's called "false contouring."
- Some digital signal sources (satellite, certain DVD releases) have an inadequate amount of video data, causing artifacts.

Bright objects rapidly moving against dark backgrounds show multi-colored flicker or trailing.
- This problem is inherent in one-chip DLP projectors using a three- or four-segment color wheel. Six-segment color wheels are less subject to it.

Hey, this isn't HDTV!
- SDTV and EDTV sets are not high-definition.
- HDTV available only via broadcast, cable, satellite HD-DVR, D-VHS (not regular VHS, DVD, LD).
- HDTV broadcasts still rare; most DTV programming is not high-definition.

Speakers

Getting your speakers to sound right has a lot to do with the mysterious arts of component matching, cable matching, and acoustic tweaking. This may take some work.

Speakers don't sound as good as they did in the store.
- Your room is acoustically different from showroom.
- Dealer may have used better amp (big difference)…
- …or better cable (slight difference).
- Speakers require a few days (or months) of break-in to achieve best performance.

Sound is compressed. Amp runs hot or shuts down.
- Bad news—impedance mismatch may require change of speakers

or amp.

Bass is weak. Sound seems hollow, disembodied.
- One or more speakers may be miswired "out of phase." Check speaker at both speaker and amp (red to red, black to black).

Overall sound is OK except for bass—too weak or strong.
- Move main speakers closer to walls, corners, floor to acoustically reinforce bass.
- Move main speakers away from walls, corners, floor to acoustically reduce bass.
- Do not block ports or vents in back of speakers—get them off shelves.

Sound is abrasive and muddy.
- Deaden room acoustics by adding absorptive or diffusive elements such as rugs, curtains, soft upholstered furniture, music/video shelving, and other elements to keep sounds from bouncing off hard surfaces, especially floors and side walls.

Sound is vague. Can't seem to get the lyrics.
- Position tweeters at eye/ear level.
- Toe in speakers toward listening position to increase proportion of direct sound.
- Smaller speakers usually produce clearer sound when mounted on stands a couple of feet or more from the wall.

A buzzing or rattling sound is audible.
- The metal baskets holding the drivers may be loose. Tighten all screws around the drivers (and be careful not to let the screwdriver slip and puncture a driver). The 20Hz-20kHz "sweep tone" found on many test CDs and DVDs will reveal this problem.
- The unwanted vibration may occur elsewhere in the room. Check ceiling tiles, light fixtures, doors, windows, moldings, bric-a-brac, etc. Don't leave things sitting on top of speakers.

Surround receivers & components

Here, at your system's heart, it's easy for a missed connection to foul up the works. After that the most common problem is signal routing—in other words, finding what button or menu will send video and audio signals where you want them to go.

No picture.
- See is TV is on.
- See if receiver is set to desired source.
- See if TV is set to appropriate video input.
- Check video connection from source to receiver.
- Check video connection from receiver to TV.
- Check video setting in receiver's menu.
- Match video connection from source to receiver, receiver to TV.

No sound, or sound in wrong channels.
- Check receiver volume/mute.
- See if receiver is set to desired source.
- Check audio connection from source to receiver.
- Check audio connection from receiver to speakers.
- Check audio setting in receiver's input menu.
- Make sure tape monitor circuit is not engaged.
- When playing DVD, select appropriate audio setting from output menu.
- When playing any surround format, make sure your equipment handles it.
- Protection circuit may be activated. Stop playback, lower volume, and power down receiver to reset it.

No sound in front left/right speakers.
- Select speaker set A or B (speaker A/B switch may be set "off" at factory).

No sound in front center and rear left/right speakers.
- Activate a surround mode.
- When playing two-channel Dolby Digital mixdown on DVD, set receiver to Pro-Logic.
- Calibrate surround processor for center and rear channel levels.

A loud hum is audible.
- If using turntable, attach ground wire to ground terminal.
- Check speaker cables for crossed wires.

- Unplug other components one by one to find source of hum; use power-line conditioner to isolate it.

Howling noise audible at high volume.
- Turntable/speaker feedback—move, isolate turntable.
- Kurt was really passionate that day.

Distorted/weak sound on LPs.
- Clean record and stylus.
- Use correct cartridge (most receivers don't do moving-coil).
- Match settings of stylus pressure and anti-skating.

Distorted/weak sound, receiver runs hot on all sources.
- Speakers present too big a load to receiver. Get a higher-powered receiver or higher-impedance speakers.

Sound is hollow, disembodied, or lacking bass.
- Phase reversed—match speaker connections red to red, black to black.

One channel weak or intermittent.
- Check audio cables for damage.
- Cassette deck heads need cleaning.

FM reception noisy.
- Move antenna to different position.
- Try longer/shorter antenna.
- Switch to mono to reduce noise.

DVD players

Usually a setup wizard will guide you through the necessary initial settings. You can always change them in the disc player's setup menu.

Disc does not play.
- Condensation may affect player or disc. Let stand 1-2 hours.
- Player may not be compatible with DVD-Audio, CD-R, CD-RW, MP3, or other formats.
- Clean disc.
- Play it right side up.

237

- Try another disc.

Some buttons don't work.
- Some operations are disabled for some discs. For example, very few players can fast-forward past the copyright warning.

No picture.
- See if TV is on.
- See if receiver is set to "DVD" input.
- See if TV is set to appropriate video input.
- Check video connection from DVD to receiver.
- Check video connection from receiver to TV.
- Check video connection from DVD to TV.
- Use same video connection (S-video, etc.) throughout signal chain.
- Most receivers will not convert one type of video connection to another..

Distorted picture.
- Check disc for fingerprints or scratches.
- Disc mastered with too much video compression—nothing you can do.
- Disc for wrong region, does not match television standard.

Frozen picture or missing footage.
- DVD is scratched or dirty. Try cleaning with a soft cloth. Wipe gently from hole outward, not with circular motion.

No sound.
- Check TV or receiver volume/mute.
- See if receiver is set to "DVD" input.
- See if TV is set to appropriate audio/video input.
- Check audio connection from DVD to receiver.
- Check audio connection from receiver to speakers.
- Check audio connection from DVD to TV.
- When playing DTS, enable DTS output from player's menu.
- When playing any surround format, make sure your equipment handles it.

No audio from digital output.
- Check digital connection between player and receiver.

- Enable that digital output from player menu.
- Enable that digital input from receiver menu.

No subtitles displayed.
- Turn on subtitles function.
- Disc may not have subtitles.

Audio (or subtitle) language not switchable.
- Disc may not have multiple languages or subtitles.
- Operation may have to be performed from opening menu.

Angle not switchable.
- Disc may not contain multiple angles (most don't).

Picture not proper shape—objects appear distorted in shape.
- Try version of movie on other side of 2-sided disc.
- Select 4:3 or 16:9 from DTV menu.
- Select "4:3 letterbox" mode from DVD-player menu.
- Some DTVs automatically go into anamorphic mode with 480p input— again, try 4:3 letterbox mode (and suffer loss of resolution, sigh).

Satellite receivers

Since there are two competing satellite-TV formats, the generic advice below will not cover format-specific problems. Also consult the trouble-shooting section in the manual, and the DirecTV (directv.com), RCA (rca.com), or Dish Network (dishnetwork.com) websites, among other places.

No service or signal.
- System requires access card.
- Card must not be inserted upside-down or backwards.

No picture.
- Check signal strength.
- Use RG-6 (not RG-59) cable between dish and receiver.
- Remove conventional cable-TV-type signal splitters from signal path.
- Check for obstacles in signal path.

239

- Check for moisture in connections.
- Level dish antenna.
- Check elevation (up/down) and azimuth (side/side) positioning.
- Check connection between satellite receiver and TV.
- Check if TV or a/v receiver is set to correct input.

No picture or frozen/intermittent/messy picture.
- Inclement weather may cause "rainfade."
- Wet snow may accumulate in dish.

Not all channels displayed in program guide menu.
- Past or future programs cannot be displayed.
- You may be looking at a "favorites" list.

Oops, forgot the password.
- Call customer service with your name, address, phone, serial number, and PIN.
- If you're that absentminded, write it down, or use something easier to recall.

Antennas

There are two main ways to foul up a TV antenna. One, choose the wrong kind. Two, install it in the wrong place.

Signal is absent.
- Check all connections.
- Center conductor of RF cable may be bent to one side.
- If antenna is active (powered), make sure power is on.
- DTV (digital) signal strength may be too low for reception.
- Check all switching (such as video recorder's TV/VCR switch).

Signal is weak.
- Raise antenna to higher position to raise signal strength.
- Attic placement severely reduces signal strength; move antenna to roof.
- Aim antenna toward transmitter location (even if it's "omnidirectional").
- Reposition antenna to minimize obstructions such as nearby buildings or towers.

- Remove unnecessary signal splitters. Don't split signal more than needed.
- Don't use 6- or 8-way splitter when 2-way splitter will serve.
- Add amplifier (or use amplified antenna) to boost signal, especially with long cable runs.
- Larger antenna, or different type of antenna, may be needed.
- Be realistic about limits of transmission distance and terrain.

TV channels 2-6 are noisier than others.
- Minimize electrical interference (if possible) by moving antenna away from power lines.
- Look for small breaks in antenna cable (effects are more visible on lower channels).
- Try motorized antenna to shift antenna position as needed.
- These "lowband" channels may require a larger antenna.

Multiple images, or "ghosts," appear on analog channels.
- Move antenna higher and/or away from objects that reflect and fragment signal.
- Try a more "directional" antenna, which is more immune from multipath distortion.

Digital channels display a pixellated, artifact-ridden picture.
- Run the automatic channel scan from the TV's or tuner's setup menu to catch up with modifications made by broadcast stations.

UHF (analog or digital) channels are weak.
- Fold down UHF elements at base of outdoor TV antenna.

FM signal is weak and noisy.
- Switch from stereo to monaural—mono signals are always stronger.

Videocassette recorders

VCRs are not as hard to use as the late-night comedians claim. The terrors of the VCR have been talk-show comedian staples for decades but they are frequently overstated.

241

Tape runs, but no picture.
- Check connection from VCR output to receiver.
- Check connection from receiver to TV.
- Check TV/VCR switch.
- If connecting to TV via RF, set same channel (3 or 4) at both VCR (see back-panel switch) and TV (channel selector).

Noise appears during picture search.
- Two-head VCRs do not perform this function as cleanly as four-head VCRs.

Picture jumping, full of streaks, or interrupted.
- VCR having difficulty tracking damaged tape. In extreme cases, eject tape to protect VCR. This often happens with rental tapes.
- Some VCRs allow manual adjustment of tape tracking, though most automate this feature, placing it beyond user control.

Picture is wobbly, sparkly, or herringboned.
- Nearby cell phones, PCs, hair dryers, or other appliances may cause interference.

Stereo hi-fi VCR plays mono low-fi audio.
- Older movies do not have stereo soundtracks.
- Older, pre-hi-fi tapes may have mono soundtracks.
- Older tapes may have linear stereo (not hi-fi stereo) soundtracks. Linear stereo playback capability is now rare.

Audio is full of gaps, while VCR flips between stereo and mono.
- VCR is switching between weak hi-fi soundtrack and mono soundtrack. Some VCRs allow you to switch over to mono, but those that automatically recognize hi-fi soundtracks may not allow this adjustment.

Setup menus appear each time VCR is turned on.
- Plug VCR into unswitched outlet—it needs constant source of power.

Recording does not start.
- No cassette loaded.
- Cassette has recording-protection tab removed.

Recording of broadcast does not work.
- Cable/satellite box must be set to correct channel. Only a VCR with cable/satellite box control can do this; otherwise, you must change channel yourself.
- VCR set to auxiliary mode (records from line inputs, not from broadcast tuner).
- VCR Plus+ code must be entered accurately.
- VCR Plus+ must be set up correctly when first used.
- Recording time is limited, depending on tape length and recording speed.

Readout does not advance with tape.
- Readout does not operate with blank tape, only with recorded material.

Clock is off by one hour.
- Turn on daylight savings mode in setup menu.

First few minutes of recordings are missing.
- Reset VCR clock a few minutes earlier.

Program guide is missing channels.
- You cannot see entries for the past or far into the future.
- You may be looking at a favorite channels list.

Remote controls

Incorrect programming is the most common problem with remotes—before you assume the unit is broken, call the technical support line.

No response from components or remote indicator lights.
- Inspect batteries for correct polarity (+/-).
- Batteries too weak to program or use remote. Replace batteries.
- Button jammed. Remove batteries, press all buttons, put batteries back in.

Indicators light, but components don't respond to commands.
- Aim remote directly at component.
- Aim remote at a less extreme angle.

243

- Remote must be close enough to component (15 feet is common).
- Make sure remote is set to command the right component brand/model.
- Make sure remote is set to correct device mode (TV, VCR, cable, CD, etc.).
- Make sure remote is set for use, not for changing components or learning commands.
- In macro mode, aim remote at component for full duration of macro.
- Component may lack ability to receive remote commands (check front panel for IR window).

Component responds to some commands, but not others.
- "Partial code"—remote is preprogrammed for a slightly different product. Look up correct model/code and reprogram remote.

Preprogrammed remote can't control a component.
- Hit all required commands before and after entering three-digit code.
- Check model number of component against list of supported models in remote manual.
- Older or rarer components may not be represented in remote's code library (try a learning remote).
- Even if a component is listed, not all of its functions may be programmable.

Learning remote will not accept new codes.
- Two or three tries may be necessary to get the remote to accept a code.
- When transferring codes, use fresh batteries in both remotes.
- Inspect batteries for correct polarity (+/-) in both remotes.
- Align original remote and learning remotes at correct distance (see manual).
- It may be necessary to put a wedge under one remote to align it with the other.
- Use a tabletop or other stable surface when transferring remote commands.
- Fire original remote for time specified by manual.
- Keep an eye on the indicators on both remotes (see manual).
- Make sure learning remote is in programming (not use) mode.

- Make sure learning remote is in correct device mode (TV, VCR, etc.).
- Hit all required commands before and after transferring codes.
- Learning remote holds fixed number of codes. Full? Eliminate nonessential codes.
- Other infrared devices in room may interfere. De-activate them.
- Direct sunlight or fluorescent lighting may interfere with transfer.
- In rare cases, some codes will not transfer properly. Use original remote.

Remote will not change TV channels.
- If original remote required "enter" command, hit "enter" on universal remote.

VCR will not go into recording mode.
- Remote or VCR may require you to hit the "record" button twice.
- Remote or VCR may require you to hit the "record" and "play" buttons simultaneously.

Can't program TV/VCR codes.
- Try the VCR codes for that brand.
- Programming of both TV and VCR codes may be required.

Touchscreen does not work.
- If screen is blank and unlit, gently tap screen to turn on remote.
- It may be necessary to recalibrate the touchscreen for your touch.
- Wrong command? A large fingertip may activate neighboring buttons.
- Use gentler touch.
- Display dim? Adjust contrast (especially after changing batteries).

Power-line accessories

Problems here should be taken with an extra measure of seriousness. There may be issues concerning safety or your system's performance.

Equipment does not turn on.
- Check all connections.
- Switch on power-line accessory. Keep it switched on at all times to feed components operating in standby mode.

- Reset circuitbreaker, or change fuse, and do not exceed product's power-handling capabilities (see manual).
- Equipment may be damaged (see below).

System does not work—one or more components fried.
- If your system was properly connected, get in touch with the surge-suppressor manufacturer. You may be entitled to replacement gear plus a new surge suppressor.

Indicators unlit or lit with wrong color.
- Be certain power-line accessory is connected to 3-prong outlet. The third prong is "ground." You may use an adapter that attaches to the screw holding the outlet plate in place (the screw should be attached to ground). But do not cheat by using an adapter that is not connected to ground—a safety hazard or performance problem may result.
- This may be a warning that polarity is incorrect, or that ground is missing or improperly connected. These things could pose deadly hazards and must be taken seriously. Get a licensed electrician to check and repair defective outlet.

Racks, stands, & mounts

Assembling them correctly is half the challenge. The other half is to correctly position your components for best performance.

Help, the parts don't fit together!
- Chill.
- Spread the parts out on the floor, with the manual.
- Try to visualize how things fit together.
- If woodscrews or shelf hardware don't fit, loosen holes with an awl, slowly and gradually.

Cables behind your rack are a mess.
- Use rack's cable management feature (if provided).
- Use plastic twist-ties to attach cables to vertical support.
- Do not group power cords with other cables (it causes interference).

Rack on wheels flexes when you move it.
- Always push or tug from bottom of rack, never from top.
- Move heaviest components to bottom shelves; don't stack components on top.
- Screw vertical supports together as tightly as possible.
- It may not be practical to move a heavily laden rack.

Newly relocated receiver/amp keeps shutting down.
- Power amps (including receivers) run hot, need space for ventilation.
- If possible, move amp to top of rack, with nothing overhead.
- Remove any components sitting on top of amp.
- Leave at least three inches of space above the amp and an inch or two at the sides.

Other newly relocated components don't sound as good.
- Keep them on separate shelves to allow proper ventilation …
- … to minimize electromagnetic or radio-frequency interference …
- … and to minimize mechanical resonances that might impair performance.
- Also keep power and interconnect cords as separate as you can.
- Relocate FM antenna to avoid interference from metal objects.

Newly relocated speakers don't sound as good.
- Fill hollow speaker stands with sand to reduce resonance.
- Attach spikes to base, coupling stand to floor.

Speaker wobbles on hinge, cannot aim in preferred direction.
- Tighten the screw that holds the hinge.
- Strengthen the hinge with a dab of glue.
- Best solution is mount rated for several times speaker's weight.

Manufacturer support

It happens in the best- and worst-run home theaters alike. Sooner or later, a component goes on the fritz. Then, in a scenario often more fraught with emotion than the works of Tennessee Williams, the hapless consumer must depend upon the kindness of strangers who work in manufacturer support departments.

You may well be about to spend a lot of time on the phone—preferably a headset phone, with some good music going in the background, assuming you still have a working audio rig. But this isn't the end of the world. And while you may feel helpless when you first realize something has gone wrong, there are ways to empower yourself, ways to make this little episode as painless as possible. Here, then, are some tips on how to survive a journey into the heart of tech support.

Be nice. This one comes first because it's the single most important thing about dealing with tech support people. No matter how upset you may be, it's important to realize that they are not responsible for the product going wrong (though you may well be). Nor are they responsible for the length of time you were kept on hold—they don't determine staffing levels, their bosses do. However, if you mind your manners, they can often (though not invariably) help make things right again. They are not your personal enemies. Their job is to help you, and they know it. Remember that—regardless of how long you were kept on hold. If you're calling a small high-end audio company, you might even find yourself speaking to the CEO!

Check the obvious stuff. Yeah, some of this may seem silly, but: Is the thing plugged in? If it's not responding to remote commands, are there batteries in the remote? (A cheap battery tester, available at any RadioShack, is an indispensable part of any home theater.) If it's winter and there's a lot of dry heat and static electricity in your home, have you tried powering down (or even unplugging) the unit to reset its frazzled electronic brain? Eliminate the obvious before you start making phone calls.

Read the manual before you call. Nearly all products come with manuals and nearly all manuals have a "troubleshooting" section. (So does this book—see the chapter on "Problem solving.") Troubleshooting guides will mention the obvious things above and a few things you may not have thought of. When you've eliminated those possibilities, give the rest of the manual a thorough read—especially if you've never done it before—to make sure you haven't simply misunderstood how the product works. Many of the calls tech support personnel receive are from people who haven't done their homework and just don't know which button to press.

Don't pop the lid. For heaven's sake, don't try to do surgery on the product yourself. If it's within warranty, your tampering will invalidate the warranty. Even if that's not the case, you could do more harm than good—perhaps to yourself. Electric shock is always a hazard even when the product has been unplugged. Its power supply and power capacitors, whose job is to store current, may still have electricity inside them. Unless you have either a degree in electrical engineering or a death wish, leave technology to the technicians.

Ask the dealer for help. If you've just bought the product, don't be a hero—exchange it. Even if that's not possible, your dealer may still be willing to provide some support if the component is a high-end product. Specialty dealers get those big markups precisely because they provide better service than the mass marketers. High-end manufacturers often choose dealers who are willing and able to provide superior support services.

Consider alternate modes of support. If the problem isn't urgent, get in touch with the manufacturer by email. You'll have to wait longer for a reply but at least you won't have to chew your fingernails off while being kept on hold. Also take a look at the manufacturer's website. In addition to the email address, the site may include other forms of support, such as FAQ (frequently asked questions) files, service-center contacts, online manuals (printable as exact replicas of the originals in the Adobe Acrobat format), and other helpful stuff. Manufacturer websites vary in scope and quality but it's never a bad idea to do a little homework there as well.

Keep the number handy. Keep a paper file of all the manuals for all your home theater components and other appliances. The tech support number may be buried in the manual, or it may be printed on a small piece of paper enclosed with the product. Don't allow yourself to misplace that documentation—and certainly don't throw it away!

Know when to call. Most consumer electronics manufacturers do not provide 24-hour, 7-day customer service. Most provide support during business hours—which may be in the eastern, central, or western time zones—five days per week. Some have longer hours than others, though, and some are open weekends. The manufacturer's website may list the hours when support is available. If not, just call, but don't be surprised if you get a recording during non-business hours. At least the recording should tell you when to call back.

Know your model number. One of the most common mistakes people make when calling tech support is not knowing the model number of the component they're calling about. DUH! Have the manual handy, and jot the serial number on the back page—you should be able to find it on the back panel or sometimes on the carton. By the way, keep the carton in storage for as long as possible, since you may need it someday to ship out a troubled component for service.

It's worth saying again—be nice! Customer service people do not have cushy jobs. They are under stress. They are not sitting in cozy private offices for the most part. They are generally in tiny cubicles, or large bullpens, wearing headset phones, dealing with cheesed-off or just plain confused people all day, surrounded by other tech support folks going through exactly the same travails. Those who don't have a union to protect them may well wish they did (and I support them). If the person you're speaking to seems nice, be advised that it takes a huge effort. If not, just be glad that job isn't yours. If all goes well the tech support person may be able to help you. If you feel your problem was not dealt with in a reasonable manner, remember that the next time you consider buying another of that manufacturer's products. As a consumer you always have the right to take your business elsewhere.

Hiring a custom installer

While the main thrust of this book is to help the do-it-yourself home theater buff, some installations will require professional assistance.

Do you need an installer? Pulling cable behind walls isn't rocket science—a local electrician can probably do the job for you. (Be sure to choose fireproof cables.) On the other hand, if you need to install a projector mount, hang a screen, or map out a multi-zone audio system, you probably will need a custom installer. Anything involving a tube-based projector automatically requires a knowledgable technician to ceiling-mount the heavy object and adjust it for best performance. Poking holes in the wall to mount a heavy plasma display—while minimizing violence to the wall—is one of many things an experienced installer can do better than an average civilian. Chain-store technicians can install satellite dishes and antennas but whether they install them well, to maximize signal strength, is another matter altogether. A good installer will also anticipate ease-of-use issues and set up your system so that any member of the family can operate it.

Consult trade groups for certification and referrals. Over the past few decades a whole industry has grown up around custom installation. Most of the better installers are certified by one or more organizations such as CEDIA (the Custom Electronic Design & Installation Association, cedia.org) or, for video, ISF (the Imaging Science Foundation, imagingscience.com). These organizations can also provide referrals to qualified custom installers in your area. Architects, interior designers, and cabinet makers also may be able to provide referrals.

Do your homework. To get off to a good start, make sure that the company is not only certified by at least one of the associations above but also fully insured and licensed to operate in the state where it does business. Speak to your local Better Business Bureau and state department of consumer protection to check whether a prospective installer has attracted complaints. Find out how long the individual or company has been in business. And, of course, obtain several references and check each one. Someone who has spent years installing car stereos or burglar alarms is not necessarily an expert on home theater or multi-zone audio. If the installer can't provide contact information for several projects comparable to yours, look elsewhere.

Verify manufacturer authorization. If you are interested in specific products or brands, make sure the installer has been authorized by the manufacturers. Manufacturers and/or their websites should be able to point you toward authorized retailers. Don't settle for a dealer who gives you an equivocal response—either he's authorized or he's not. If he's not you may end up with grey-market goods that lack proper warranties. The right installer/retailer is the one who provides superior service when a product goes awry, as even high-end products sometimes do. Smart manufacturers are scrupulous about who they allow (and do not allow) to sell and service their products.

Pay a visit. A good installer does not necessarily have to be associated with a large store but should present a businesslike appearance. "A visit to the installer's office is worth thousands of dollars," according to Mitchell Klein, president of Media Systems (Boston, West Palm Beach) and former president of CEDIA. Is the person neat and well-organized? Are your calls answered by a human being or at least returned promptly? This is who you will be calling when something goes wrong.

Talk, listen, and observe. The instant you begin speaking with a custom installer, information is being exchanged, and it moves in both directions. Of course you should write down your questions before the meeting and ask them. And your questions should receive satisfactory answers. But don't try to do all the talking. A good installer will ask about your goals, space, limitations, and budget. Having determined your needs and expectations, he will provide options and price quotes. Recognize that the cheapest quote may not result in the best service.

Get it in writing. When you're ready to proceed, get every detail of the job in writing. The proposal may include things like wiring, schematics, elevations, documentation, and project meetings. Try to anticipate cost overruns and upgrade issues. Unexpected problems may come up, especially in older homes that were not built with electronic systems in mind, though newer ones may be easier, especially those that have been pre-wired by the builder.

Be a good collaborator. Whether you need to be firm or flexible about arising problems is your own judgment call. If you've chosen a reputable installer, and find the person trustworthy, try not to be too hard-nosed. This is a relationship like any other. Treat a good installer as a collaborator and expected to be treated the same in return. A wisely chosen and intelligently handled custom installer should prove to be the best ally you could ever have imagined.

Acknowledgments

This is the 10th edition of an annually updated resource. Home theater does not stand still. Its underlying technologies, big-screen television and surround sound, are rapidly evolving. *Practical Home Theater* will grow and change along with it. Please feel free to contribute comments and ideas for changes in future editions via email—you can find the address, which may change from time to time, on the contact page at quietriver-press.com. It may not be possible for me to provide Q&A service but whatever you have to say will be read with interest.

Even before the first edition of this book came out in 2002 I had planned annual updates. Over the years, parts of the book have been substantially reorganized and rewritten, especially the television and surround sound chapters. This year's edition is the first to include a chapter on 3DTV. Two chapters formerly devoted to digital video recorders and digital audio servers have been consolidated into a single chapter on DVRs, streamers, and servers—partly to recognize the overlap among these categories, and partly to better accommodate new content. HDMI continually brings new versions and surprises. And as always, there has been a drumbeat of other small changes.

Many thanks to the expert readers who were kind enough to read and comment on portions of earlier drafts. Special thanks to Brent Butterworth (then of Dolby Laboratories, now contributor to *Sound & Vision*) for giving the audio coverage the benefit of his world-class expertise, and to Lancelot Braithwaite (former technical editor of *Video Magazine*, now with *Widescreen Review*) for helping me better understand the technology of television. They did the heaviest lifting and I can never thank them enough. Thanks to Jeff Samuels and Bill Schindler of Panasonic for helping with the 3DTV chapter—please note that, technical details aside, my skeptical take on the subject is entirely my own. I'm also grateful to Michael Guillory of Texas Instruments for help with DLP technology and to Scott Wilkinson for recommending Blu-ray test discs.

I couldn't have survived the chapter on "Understanding surround standards" without the help of Craig Eggers (Dolby Labs) and Don Dixon (DTS). Thanks again to Craig Eggers (then of Toshiba) and Len Schneider (Technicom) for reading the chapter on DVD, and to Gregory Thagard (Warner Bros.) and Mikhail Tsinberg (Key Digital) for helping me better understand progressive-scan DVD. I'm also grateful to Brian Dietz and Neal Goldberg (NCTA), Bob Perry (then of Mitsubishi, now of Panasonic), John Taylor (LG/Zenith), and Mike Schwartz (CableLabs) for bringing me up to speed on digital cable readiness.

Joel Silver (Imaging Science Foundation) and Marty Zanfino (then of Mitsubishi) also shared their video expertise. John Dall (THX) and Graham McKenna (then of Pacifico PR) helped fine-tune the details about THX. Laura Hubbard of the Consumer Electronics Association was kind enough to provide key statistics.

Please note that while these industry luminaries were kind enough to help weed out the garden, they and their employers or clients are in no way responsible for any remaining weeds. The author accepts blame for any errors. Opinions expressed between these covers are solely those of the author.

Thanks to the following folks for providing products photographed for the book: Joe Abrams (ex-Music Interface Technologies), Lisa Borman (Borman PR), Steven Hill (Straight Wire), Daniel Graham (Monster Cable), Joe Perfito (Tributaries), Stan Pinkwas (J.B. Stanton Communications), Phil Raymondo (Esoteric), Roger Skoff (XLO), Bryan Stanton (J.B. Stanton Communications), and Lois Whitman (HWH PR).

I'm deeply grateful to all of the editors who have supported my writing career over the past year with steadfast support, patience, kindness, and good editing. They currently include Shane Buettner and Claire Lloyd of *Home Theater*, as well as Doug Garr, who inspired me to become a writer. Like any working writer, I am indebted to more sources of collegial support than can be listed here.

Mark Fleischmann
New York, October 2010

About the Author

Mark Fleischmann is a New York-based writer specializing in technology and the arts. Currently serving as audio editor of *Home Theater*, he reviews audio/video gear, writes the magazine's AV News column, and contributes daily news stories and blogs to hometheatermag.com. He was a co-founder of *etown.com* (1995-2001), in its time the world's most widely read consumer electronics publication, and was its first and longest-serving editor-in-chief. Mark was audio critic of *Rolling Stone* for nine years and has written columns on audio/video hardware for *Digital-Trends.com*, *Premiere*, and *The Village Voice*. His column on home theater ran for 15 years in *Audio Video Interiors*, making it the longest-running column ever written on the subject. His writing has appeared in *Amtrak Express*, *Bloomberg Personal Finance*, *Business Week*, *Cargo*, *c|net*, *CrutchfieldAdvisor.com*, *Custom Retailer*, *Details*, *E-Gear*, *Elle Decor*, *Harper's Bazaar*, *Help.com*, *The Men's Journal*, *Musician*, *Penthouse*, *Popular Science*, *The Robb Report*, *Spin*, *Stereo Review*, *The Stereophile Guide to Home Theater*, *Sync*, *Ultimate AV*, *The Washington Post*, and many other publications. Mark has served both as a movie/video critic (for *Entertainment Weekly*, *Newsday*, *Video*) and as a music critic (*Musician*, *Spin*, *Trouser Press*). A former senior editor of *Video* (1980-86), he edited the magazine's video programming section, as well as a record collector's magazine for *Trouser Press*. He is the author of another book, *Happy Pig's Hot 100 New York Restaurants*, converted to a website at happypig100.com. His other web destinations include quietriverpress.com (book publishing) and mfwriter.com (career survey). Mark still plays his LPs and still feels silly writing about himself in the third person.

Index

Breinigsville, PA USA
09 March 2011
257289BV00002B/245/P